Discovering

Somers

Isles

A Guide to Bermuda's History
1500 - 1615

By John M. Archer

PUBLISHED BY MAURY BOOKS

Printed and Bound in the United States of America

ISBN: 978-0-9963455-5-2

Cover image from "The Gust,"

Painting by Willem van de Velde, the Younger, circa 1680

Cover design by the Author

Dedicated to

the men and women of the "Sea Venture";

their courage and determination helped

shape two countries.

Contents

Images

Images (cont.):

Image Credits:

Pg. 29: Image from ship's model on display at Dorset County Museum, Dorsetshire, Dorset, UK. Original photo by user Musphot on Wikimedia Commons

Other:

Lefroy, J.H. <u>*Discovery and Early Settlement of the Bermudas*</u>

LC   Library of Congress

LookBermuda   Look Bermuda Destination Media Services, Bermuda

Whymper. Frederick, <u>*The Sea: Its Stirring Story*</u>

WP   Wikipedia (Public domain image)

WC   Wikipedia Common

Unless otherwise credited, images are from author's collection

Notes to the Reader:

Spelling:

Quotes from the seventeenth-century accounts used in this book often contained what we would see as misspellings and odd sentence structure. For ease of reading, the author has "modernized" this content when necessary, but without changing the meaning of the passage.

Dates:

This history primarily concerns English colonists and mariners of the early 1600's and uses what was termed the "Old" or Julian calendar. Although decreed by Pope Gregory in 1582, the Gregorian calendar in use today was not observed in England and the colonies until 1752. Technically, the dates used here are about ten days behind our current calendar. For simplicity sake, the anniversaries of these dates are still observed in Bermuda on the original Julian dates.

Acknowledgements

Unlike previous research projects close to home, creating a tour of an area at a distance of eight hundred miles and four hundred years had its challenges. I have been fortunate enough to have had the help of a number of kind and talented individuals.

I am especially grateful to Dr. Edward Harris and Jane Downing at the National Museum of Bermuda. Their assistance and kind words along the way were invaluable.

Three images in this book appear courtesy of artist Christopher Grimes. Many thanks to Mr. Grimes for his generosity - and my apologies for presenting his wonderful paintings in this small black and white format. My appreciation also goes to Jennifer Gray and Dorte Horsfield of the Bermuda National Trust for their help with the image of Sir George Somers' map.

Of course, my greatest thanks goes to my wife Darlene. Her patience and proof-reading skills not only saw this project through, but in this case, it was she that convinced me to take that first voyage to Bermuda.

Lastly, my appreciation goes to the people of Bermuda itself; in ways great and small, their warmth and character inspired me to write this book about their home.

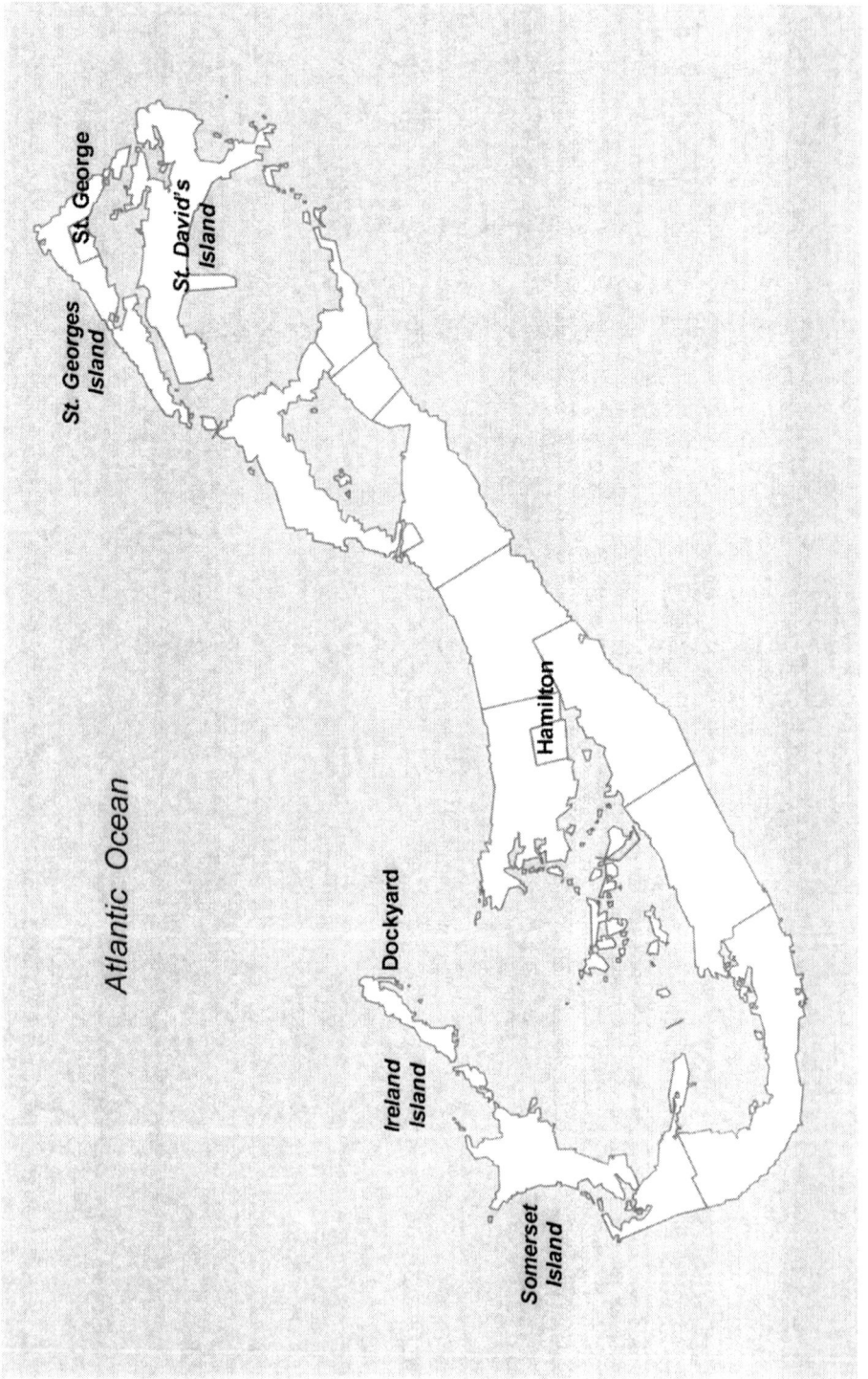

Atlantic Ocean

St. Georges
Island

St. George

St. David's
Island

Dockyard

Ireland
Island

Somerset
Island

Hamilton

No excuse seems necessary for attempting the preparation of an adequate guide-book for what is undoubtedly one of the most delightful of all the little corners of the world.

- Fremont Rider, Bermuda: A Guide Book for Travelers, 1922

Introduction

As so well expressed in Fremont Rider's quote, in a country as uniquely beautiful as Bermuda, little excuse seems necessary to write a guide, or further, spend time exploring how the islands were first settled. And perhaps the best way to understand that tale is to leave the history books behind and walk the same ground where those events occurred. However, for those wishing to explore Bermuda's origins, particularly visitors with little local knowledge, there has not been an interpretive guide to the sites that played a key role in the settlement of the islands.

This book is an effort to remedy that oversight; it is not intended as a comprehensive history of the islands; Bermuda's 400 years of social, economic, and political growing pains have been well documented in several works (some of those titles can be found in the *Bibliography*). Instead, this guide is intended to allow the reader, whether in Bermuda or not, to follow the discovery of what were first known as "Somers Isles" for themselves.

The text begins with a look at the islands' sinister reputation in the 1500's, and some of the first victims of the still-treacherous reefs that surround Bermuda. The next section uses a tour format and period accounts, investigating the fateful voyage and wreck of the English ship *Sea Venture,* exploring the area where the survivors landed, their struggle to survive, and the early days of the settlement on those once desolate "Devil's Isles."

This is no easy task; the average post-Elizabethan settler was not literate (or could read but not write), so the information here relies on relatively few accounts as well as oral traditions. And, to help understand their perspective, readers may find themselves relatively far afield—for instance, vessels and seafaring of the period are briefly discussed to help appreciate the hardships faced by these early travelers.

In some areas, seeing the landscape as it was in the 1600's can be a challenge as well; not surprisingly, the needs of modern Bermuda— heavily populated islands with a busy tourist economy—have necessarily altered much of the ground where this story took place. To compensate, the text is accompanied by maps and images designed to help the reader envision the area as early settlers saw it, with "seventeenth-century eyes," if you will. It is the author's hope that the reader will be rewarded with a better understanding of the unique character of the Bermuda islands, and those who settled there.

In the end, as you follow this tale of Bermuda's discovery and founding, whether visiting on-site or reading at home, bear in mind this account is only "history" in the sense it deserves remembrance; on Bermuda's shores, the survival and perseverance of those who first came there is still very much alive.

"You can go to heaven if you want. I'd rather stay in Bermuda."

- Mark Twain

The New World of 1540 – Early map by Sebastian Munster

Prologue: A New World

The Western Hemisphere of the sixteenth century was a place of seemingly unlimited promise. With Columbus' initial explorations of a "New World" in the Caribbean in the 1490's, the prospect of colonization offered England and European nations a significant opportunity to expand their influence, wealth, and respective religions.

Despite the rush to explore and settle the New World, the fact that Bermuda was discovered *at all* might seem amazing; the tiny cluster of islands sits almost in the middle of 20,000,000 square miles of the North Atlantic. But as trade routes developed to the West Indies in the 1500's, the return journey used the favorable westerly winds and currents of the Gulf Stream, bringing hundreds of ships past the islands' treacherous reefs—and leaving countless ensnared on them. Ironically, those distinctive coral reefs that made Bermuda so unapproachable contributed no small part in how the island came to be settled.

As it happened, the English were relative latecomers to colonizing the New World. In 1607, their latest settlement, christened Jamestown, was established on the Chesapeake with high hopes. However, the following year saw almost two-thirds of the colonists laid in the sandy clay shores of the James River, victims of poor survival skills, the settlement's marshy location, and a rocky relationship with the Native Powhatan tribes. Seemingly, England's only surviving colony in the Americas was doomed.

Rumors of the tragedy began to reach London and the organizers of the original voyage, the "Virginia Company." In May of 1609, a relief convoy consisting of nine vessels laden with some 600 new settlers and tons of fresh supplies set sail to relieve Jamestown. En route, the ships ran into a hurricane that scattered the fleet; battered and sinking, the flagship of the convoy was run aground by the fleet's admiral, Sir George Somers, on a cluster of islands in the middle of the Atlantic. There, the confluence of people, place, and events produced results that rippled far beyond the original intent of those involved. What follows is the extraordinary tale of how those islands came to be settled.

Part One

"The Isle of Devils"

Exploring the Atlantic Ocean in the sixteenth century was a risky affair at best. Early mariners rightfully saw the oceans as a treacherous place: in addition the more familiar hazards faced by wind-driven vessels, the seas were believed to be filled with demons and capricious spirits ready to take the unwary to a watery grave—or worse.

"Les Monstres Marins" - Sebastian Münster, 1556

After weeks at sea, any seadog would have been glad to spy land on the horizon; if it were *"The Bermudas,"* that joy would be short-lived. According to rumor, if one did manage to navigate the broad line of breakers and reefs surrounding the islands to get within earshot, unearthly cries could be heard from the shores. Little did it matter the cries came, not from the spirit world, but an indigenous bird known as the cahow. At all events, the islands' unpredictable winds would give any wizened seafarer pause—those who ignored them soon found themselves wrecked on the treacherous reefs that surrounded its shores. It is little surprise that, before long, many knew the islands simply as the "Isle of Devils." 1

1500-1603: The First Visitors

The archipelago that came to be known as Bermuda begins to appear in maritime records about 1505. According to Spanish historian Gonzales Oviedo, one Juan de Bermudez discovered the islands while sailing to Mexico, and soon after, "La Bermuda" first appears on a 1511 map of the New World. In 1515, Oviedo himself attempted to land at the east end of the islands to explore and leave hogs behind for marooned sailors, but failed to land due to the island's "contrarie winde."

In 1543, a Portuguese shipwreck left survivors stranded on the island's south shore. There they constructed a new vessel to continue to the Caribbean. It was probably these marooned mariners who left a carving on a high cliff overlooking the ocean on the island's south shore. Known today as "Portuguese Rock," an inscription containing what appear to be letters and "1543," could long be seen

Inscription from Portuguese Rock

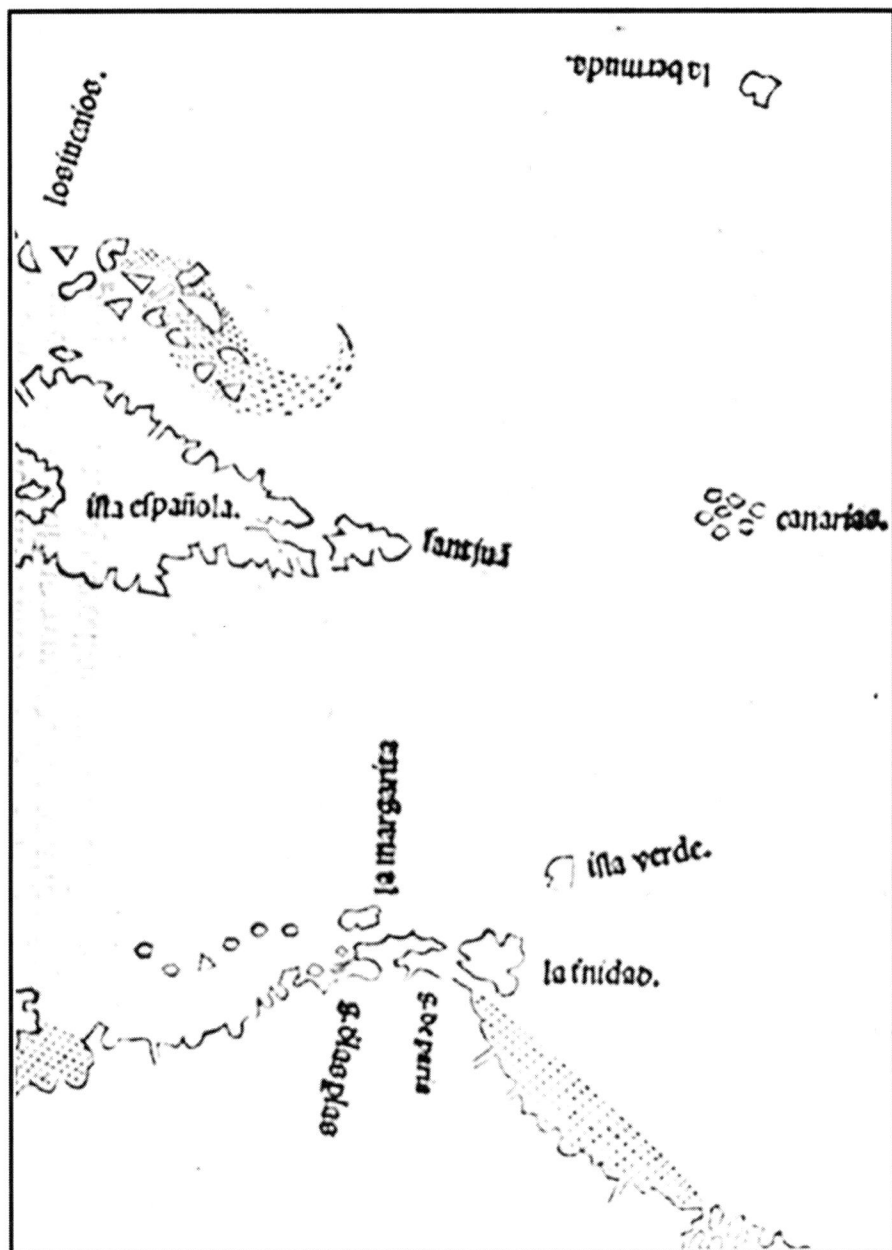

This 1511 map drawn by New World Historian Peter Martyr d'Anghiera shows "La Bermuda" for the first time (upside down at top). The Caribbean and South America appear at lower left. (LC)

at Spittal Pond Reserve in Smith's Parish. (After years of damage by erosion and vandalism, the boulder is marked today with a plaque and facsimile of the carving). 2

In addition to this inscription, a number of later accounts describe finding "crosses, peeces of Spanish monies here and there," as well as strange marks on trees, such that one islet in Great Sound came to be known as "Cross Island". At the opposite end of Bermuda, Cooper's Island was said to contain yellowwood trees bearing letters and a brass plaque, and a "tryangle" of stones. Not surprisingly, legends soon grew of Spanish treasure buried on the islands. Ironically, perhaps the most valuable treasure was the yellowwood itself; never abundant, the wood became so prized in England for its color and grain, that despite regulations forbidding its export, the original species no longer grows on the islands. 3

These "Broken Isles"

In the 1590's, another luckless sailor named Henry May was the first Englishman to set foot on Bermuda. After a number of misadventures in the Caribbean on his first ship, May was attempting to return to England aboard a French vessel. On the moonless night of December 17, 1593, the ship ran aground on the reefs north of Bermuda. In the pitch dark, it appeared to May that they had foundered on an island with, "hie cliffs." In fact, the following dawn revealed the ship was on a reef far from land. (If May's description is accurate, it is likely the ship hit the broad, pillared reef northeast of Bermuda known as North Rock, actually some eight miles from shore. (See the *Appendix* for more about this unusual reef feature). 4

At sunrise, May and twenty-six survivors rowed seven leagues—about 21 miles—aboard the ship's longboat before finding a gap in the reefs. Making landfall just before sunset, May found a landscape "divided all into broken Islands," many wooded with a variety of trees, "but Cedar is the chiefest." For sustenance, they discovered very little water, some thin hogs (no doubt left by earlier voyagers like Oviedo), but also, "a great store of fowle, fish, and tortoises." On the east end of the island, he found "very good harbours, so that a shippe of 200 tun may ride there land-locked, without any danger." (May was also the first to state the island was "as good for the fishing of pearls as any in the West Indies," a claim not borne out by later divers). 5

In an uncanny parallel to what was to follow some 15 years later,

Shipwreck on North Rock, from an 1873 engraving (F. Whymper)

using tools, sails and rigging salvaged from the wreck, the survivors built "a smalle barke of some 18 tuns," using the islands' abundant cedar. In 1594, May and his companions reached Newfoundland, and finally arriving in England that August, where May published the first close-hand description of Bermuda. 6

Soon after, in 1603 a Spanish galleon under Captain Diego Ramirez also ran aground, but managed to anchor off Bermuda for repairs. "The first night that I anchored in the bay, I sent a small boat to an inlet to look for water, but none was found," Captain Ramirez wrote, "At dusk, such a shrieking and din filled the air that fear seized us." On the dark shore, Ramirez' men were understandably afraid, particularly after one of the crew was set upon by cahows attracted by his lantern. "Only one variety of bird makes this noise, but the concerted yell is terrible...." 7

Eventually, during his stay Ramirez sketched and made detailed notes of his stay on the islands. In Great Sound, Ramirez noted an island marked with a cross; as described earlier, the marker on "Cross Island" was one of the locations thought to point to buried treasure; in all likelihood, the cross pointed to the location of something equally valuable: fresh water. (Generally known today as Cross Island, today the islet forms part of the jetty at the Royal Naval Dockyard). Ramirez forwarded his descriptions to King Phillip's court back in Spain; however, even with Bermuda's supernatural properties debunked, and the island's potential as an Atlantic way-station, the "Isle of Devils" still held no fascination for the King of Spain.

1606-14: The Virginia Company

By the beginning of the seventeenth century, despite numerous reports of the potential benefits of the Bermudas as a stop on the Atlantic trade routes, the reputation of those forbidding shores had left it a mere geographical marker. Ironically, it was England, all of whose previous colonial efforts had failed, who came to realize the importance of the islands, albeit quite by accident.

Beset by a series of economic and internal problems, not the least of which was the undeclared war with Spain, English colonial efforts initially saw a series of failures, notably the Lost Colony at Roanoke in the 1580's. For the next two decades interest in further attempts waned. However, in 1603 a number of changes occurred in England that would place Bermuda in the path of English exploration. After some 45 years as Queen, Elizabeth I died, and James I became king. Unlike Elizabeth, the new monarch was willing to support more attempts to colonize the New World.

In 1606, James chartered "The Virginia Company" to establish settlements along the east coast of North America. Representing the cities of Plymouth and London, the charters gave the Company rights to seek profit by settling "Virginia," at the time an area considered to stretch from the Carolinas on the south to New England on the north.

From the beginning, the attempts at colonization operated as a joint public-private venture. (Indeed, Italian investors partially funded Columbus' voyages to the Americas for Spain). The Virginia Company was composed of an odd mixture of investors: wealthy gentlemen, commoners, along with a good number of London's ne'er-do-wells, but all looking for a new life, property, and riches.

However, with the Company's focus on gathering the trades necessary for turning a *profit*, little attention seems to have been paid to attracting those with talents necessary for *survival*: farmers, carpenters, and fishermen.

Accordingly, the first Plymouth Company expeditions soon failed: the Spanish, protective of their New World holdings, seized one group before it even reached Virginia, and a colony established at Sagahodoc in Maine had been abandoned by 1608. All that remained for the Company's efforts was the 1607 settlement on the Chesapeake at "Jamestowne"; but those colonists were struggling as well, victims of a marshy location, poor survival skills, and a rocky relationship with the Native Powhatan tribes. Despite two more attempts to resupply the settlement, rumors of a tragedy began to reach London: England's only surviving colony in the Americas was failing.

The organizers of the Virginia Company, anxious that their latest investment succeed, quickly made plans for another relief convoy. It was Sir Thomas Smythe, the Company's treasurer and one of the leading entrepreneurs of the day, who lobbied King James to issue a second charter for the Virginia Company in 1609, paving the way to raise monies for a new convoy to rescue the colony.

The Company's "Third Supply" attempted to avoid the mistakes of previous expeditions; they would only be partially successful. To prevent the power squabbles that crippled Jamestown's leadership previously, at the outset one of the Company founders, Sir Thomas Gates, was designated deputy governor. A knighted gentleman, Gates was also a soldier who had served in the wars in the Netherlands; now responsible for perhaps the largest one-time migration to the New

Sir Thomas Gates

Sir George Somers

World, Gates could be expected to rule with a firm hand while also protecting the colonists from the Powhatan.

While at sea, Sir George Somers was to command the fleet. Also a founding member of the Company, Somers' 55 years had seen a life as privateer, naval hero, wealthy merchant, and knighted Member of Parliament. Commanding Somers' flagship, the *"Sea Venture,"* was another well-respected mariner, Captain Christopher Newport. Despite losing an arm while privateering against the Spanish in the Caribbean, Newport had spent twenty years at sea, including several voyages to Virginia; there he served on the governing council at the initial Jamestown settlement, and knew, perhaps better than any, what lay ahead.

However, similar to Jamestown, the egos of those in charge became a problem almost immediately; as described in one account, "All things being ready, because those three Captaines could not agree for place, it was concluded they should goe all in one ship," the flagship *Sea Venture.* By the end of the voyage, this surplus of dynamic leaders likely created as many problems as it solved. 8

In the meantime, the Company organized a successful campaign to woo investors to save Virginia for the glory of God and Country— and potential personal gain. By May of 1609, hundreds of "adventurers" had invested; some 600 signed on for the voyage itself: some pledging a monetary investment, while others pledged part of their lives: indentured work for the colony. Unlike the original convoy, it appears a higher percentage of the Third Supply included the trades necessary for survival: carpenters, bakers, tailors, fishermen and the like, although once again, records indicate they were joined by a large number of the unskilled from all classes of London's social strata. But in the end, all were daring enough to risk their lives and fortunes for the opportunity they saw in the "Virginia Plantation."

Logistics

With the choices and convenience of modern travel, it is easy to forget that, until relatively recently, crossing the Atlantic was a very uncertain journey. The logistics for Jamestown's Third Supply would be enormous; the convoy consisted of seven ships of varying sizes along with two pinnaces (akin to a mid-size sailing yacht) to carry 600 colonists, the foodstuffs necessary to survive a grueling eight-week trip, as well all as the supplies required to reestablish the settlement.

Below decks, stores of what was considered an adequate diet occupied much of the space: salted beef, pork, and fish, as well as bread, flour, cheese, vegetables and fruit, all stored in casks and hogsheads. Storing liquids at sea, however, was more problematic; after a few days in casks, even water became a broth of bacteria, particularly in tropical climes. Alcohol-based beverages, beer, and wine lasted longer and, not surprisingly, became a seadog's beverage of choice. Even so, these liquids would turn in time; period accounts often describe mariners lamenting their spoiled beer.

In an ocean plied by known enemies and random privateers looking for ready profits, merchant ships also needed to be well armed, and most sailed with a variety of short and long-range cannon. With no guarantee of rescue in foreign lands, in the event of damage Elizabethan vessels needed to be self-reliant, carrying all the tools and supplies needed for repairs, or if necessary, to construct a new vessel. As a result, gunpowder and various forms of ordnance shared the remaining space below decks with more prosaic items: spare sails, tools, rope, iron, and the like. It is little wonder that ships of the seventeenth century were measured by their tonnage in cargo capacity rather than their actual size.

The Sea Venture

For all that history knows of her fate, relatively little is known about the layout and appearance of Somers' flagship, the *Sea Venture*. At the time, the vessel was described as, "three-hundred tunne," rather mid-sized for her day, and marine archaeology confirms the vessel was about one hundred feet long from stem to stern.

Built about six years earlier, Somers' vessel may well have benefited from design changes instituted on the Queen's ships by some-time naval hero Sir John Hawkyns. An English merchant and shipwright, Hawkyns' designs married the defensive capability of the formidable Spanish galleon with the carrying capacity of a swift merchant vessel. Possessing lower fore and stern-castles and longer length, these "race-built" ships were more weatherly, meant to be stable and seaworthy not only in the English Channel, but on the high seas. In the event, the construction of the *Sea Venture* would more than fulfill the promise of its builders. 9

In early May, the loading of the *Sea Venture* began. Doubtless, Admiral Somers and Captain Newport were aboard from the first, preparing the ship with about twenty to thirty crewmembers. The Admiral was accompanied by an aide, Lieutenant Edward Waters;

Typical 17th-Century Merchantman (Approx. 400-ton)

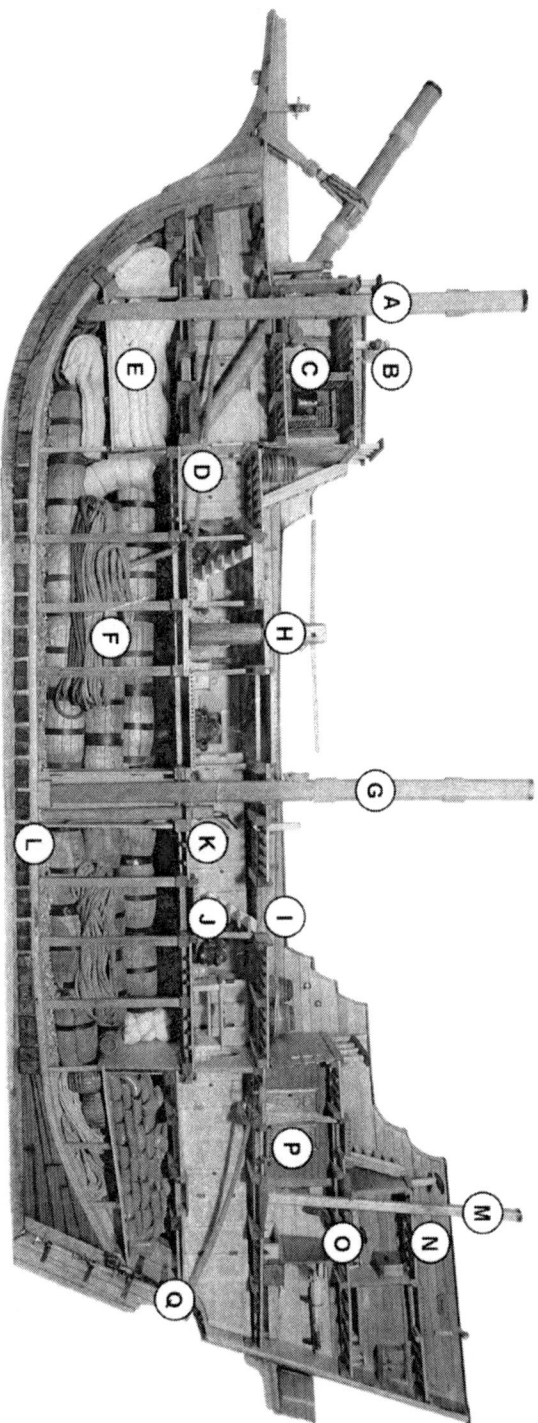

A. FORE MAST
B. FORECASTLE
C. GALLEY
D. CREW SPACE
E. SAIL LOCKER
F. STORES

G. MAIN MAST
H. CAPSTAN
I. MAIN DECK
J. GUN DECK
K. BILGE PUMP
L. BALLAST

M. MIZZEN MAST
N. STERNCASTLE
O. OFFICER QUARTERS
P. WHIPSTAFF
Q. TILLER & RUDDER

Model of "Mary and John" on display in the Dorset County Museum, Dorchester, Dorset, UK
Original photo by user Musphot on Wikimedia Commons

the lieutenant would come to play an important role in the months to follow. Others were likely trusted men that the two officers had sailed with on previous voyages—veteran mariners like coxswain Robert Walsingham, able seaman Robert Waters (no relation to Edward), shipwright Robert Frobisher, and carpenter Nicholas Bennit; most would live up to that trust, others would not. 10

On this voyage, the *Sea Venture* would carry 150 adventurers and crew, their baggage, a dog, and the lion's share of the supplies, which may have included horses and other livestock. To accommodate the colonists, the vessel's cannon were removed from the gun deck and installed above on the main deck. The end results were mixed: it allowed the former gun deck to be partitioned into the small

Possible appearance of the Sea Venture (LookBermuda)

compartments necessary to house well over 100 colonists; but with the cannon mounted above, the ship would be more top-heavy—a dangerous arrangement if high seas were encountered.

When the passengers finally started to arrive, the procession up the gangway must have been a sight to behold: the notables included another of Lyme Regis' merchants, Sylvester Jourdain; also, one William Strachey: sometime poet, theatre aficionado, friend to London's literati—and recurrently unemployed; but both men were educated and literate, later penning detailed accounts of the voyage. Key to the Company's interests, gentleman Thomas Whittingham was the cape-merchant who was to oversee the supplies, making sure the new settlement was run profitably. Another passenger, "Mistress Horton," boarded the *Sea Venture* with her maid, Elizabeth Persons; we can only guess what they hoped to find in Virginia. Despite the many sermons in London about building a Protestant Virginia, the only minister to accompany the convoy was Reverend Richard Bucke, trusting his faith would protect his wife Maria and their two daughters during the journey. After the passengers of relative privilege had boarded, those whom William Strachey would later refer to as the "common sort" made their way onto the ship, no doubt carrying all they had of value in the world. Regardless of their station in life, the voyage was about to give all 150 passengers and crew much in common.

On May 15, the convoy was making its way to Plymouth on England's southwest coast. There the ships waited most of a week; delayed by contrary southwesterly winds, they finally weighed anchor on June 2. Customarily, the route to North America called for sailing south to the Canary Islands, then taking a westerly heading to take advantage of the prevailing currents to the Caribbean, and then sailing north to reach the Virginia coast. For this voyage, Company officials had specified a new route: to skirt a potential threat from Spanish ships operating in the Indies, the convoy was to turn west earlier than usual, and find winds that would bring a more direct route to Virginia ahead of the Atlantic's summer storms.

"A dreadful storm and hideous"

For seven weeks, the accounts of the journey suggest the convoy had fair weather, and sailed "in friendly consort," scarcely losing sight of one another; these accounts belie the miserable voyage the colonists endured. Today, most travelers take for granted the comforts available on stabilized passenger ships; even in fair weather, seventeenth-century ships of sail rolled with each rise and fall of the waves; most of the passengers were seasick from the start. Even if one could stomach solid food, after a week or two the color and smell of the food from the ship's hold might not inspire confidence. And, as described earlier, the ships' drinking water likely began to turn foul as well. Even so, water of any condition was a precious commodity: bathing was out of the question.

Then, after sunset on Sunday, July 24, the previously calm seas began to rise. Monday morning saw the skies turn dark. The description written by William Strachey is telling:

The clouds gathering thick upon us and the wind singing and whistling most unusually...a dreadful storm and hideous began to blow from out the northeast... The sea swelled above the clouds and gave battle unto Heaven. It could not be said to rain: the waters like whole rivers did flood in the air...For my own part, I had been in some storms before, as well upon the coast of Barbary and Algiers...Yet all that I had ever suffered gathered together might not hold comparison with this...There was not a moment in which the sudden splitting or instant oversetting of the ship was not expected...And the manner of the sickness it lays upon the body, being so unsufferable, gives not the mind any free and quiet time to use her judgment. 11

By Tuesday, after an unimaginably terrifying night, it is easy to picture all hands hiding below decks in the gale. Although most of the sails "lay without their use," certainly Newport and Somers held their crewmen to their posts, repairing the battered ship as best they could in the midst of the gale. But they soon discovered that the *Sea Venture*

faced yet another threat, this time from onboard: the constant strain of the wind and seas on the vessel's hull had forced much of the oakum—the waterproof seal of hemp and tar—from between the wooden planks of the hull; the hold began to fill with water. Strachey continued:

> *Howbeit this was not all. It pleased God to bring a greater affliction yet upon us; for in the beginning of the storm we had received likewise a mighty leak. And the ship...was grown five foot suddenly deep with water above her ballast, and we almost drowned within whilst we sat looking when to perish from above...The most hardy mariner of them all...now began to sorrow for himself when he saw such a pond of water so suddenly broken in and which he knew could...but instantly sink him...*

> *Our governor upon the Tuesday morning...had caused the whole company (about 140, besides women) to be equally divided into three parts, and...appointed each man where to attend; and thereunto every man came duly upon his watch, took the bucket or pump for one hour, and rested another. Then men might be seen to labor, I may well say, for life; and the better sort (even our governor and admiral themselves), not refusing their turn and to spell each the other...*

As Wednesday dawned, the storm only intensified. At the mercy of the seas, and the tons of water in her hold, the *Sea Venture* began to list badly to starboard. The desperation is plain in Strachey's words:

Once so huge a sea brake upon the poop and quarter [decks] upon us as it covered our ship from stern to stem like a garment or a vast cloud; it filled her brim full for a while within, from the hatches up to the spar deck. [W]e much unrigged our ship, threw overboard much luggage, many a trunk and chest (in which I suffered no mean loss), and staved many a butt of beer, hogsheads of oil, cider, wine, and vinegar, and heaved away all our ordnance on the starboard side... [E]very four hours we quitted one hundred tons of water, and from Tuesday noon till Friday noon...

Working with tired bodies and wasted spirits three days and four nights, destitute of outward comfort and desperate of any deliverance we were much spent...without either sleep or food; for the leakage taking up all the hold, we could neither come by beer nor fresh water; fire we could keep none in the cook room to dress any meat.

If Strachey's narrative seems overwrought to our ears, fellow passenger Sylvester Jourdain described much the same desperation:

[The Sea Venture] received so much water as covered two tier of hogsheads above the ballast, that our men stood up to their middles with buckets, barricos, and kettles to bail out the water...yet the water seemed to increase than to diminish... Our men being utterly spent, tired, and disabled for longer labor, were even resolved, without any hope for their lives, to shut up the hatches and to have committed themselves to the

sea. . . . So that some [of] them having some good and comfortable waters in the ship fetcht them, and drunk one to the other, taking their last leave one of the other, until their more joyful and happy meeting in a more blessed world. 12

Doubtless with little hope left, Somers still stayed above, directing the helmsman at the whipstaff on the deck below. Then, squinting at the horizon through exhausted eyes, Sir George caught sight of a dark shoreline above the waves. Strachey later recounted:

Surely that night we must have...then perished. [But] Sir George Somers, when no man dreamed of such happiness... discovered and cried land. Indeed the morning, now three quarters spent, had won a little clearness from the days before, and it being better surveyed, the very trees were seen to move with the wind upon the shore side... 13

As the foundering vessel approached the rocky shore, the leadsman and the boatswain took to the bow to read the depth, and word was relayed to the poop deck. "By the mark, seven!" or seven fathoms: 42 feet; "A quarter less four!": about 26 feet. The *Sea Venture* was now less than a mile from the beach visible to the northwest.

Perhaps testament to Somers' and Newport's skills, or some combination of that skill and blind luck, the *Sea Venture* headed into one of the only channels between the reefs at Bermuda's East End. However, as it happened, the waves hurled the ship into a V-shaped gap in the reef; the coral cut into the hull with a grinding roar, and everything aboard came crashing forward. The still-violent surf

"The Wreck of the Sea Venture"
(From oil painting by Christopher M. Grimes)

Modern view looks toward St. Catherine's Beach. The Sea Venture struck the reefs near here.

continued to drive the stricken vessel further into the cut, but soon, the terrifying crash from the ship's hull may well have ended as suddenly as it started.

Over the sound of the waves, only the sounds of terror remained; screams, cries, and curses all quickly gave rise to a new panic: the stricken ship was still over half a mile from shore, taking on water—and the average seventeenth-century city-dweller could not swim. Captain Newport immediately ordered the crew to lower the *Sea Venture's* longboat and a second smaller craft into the choppy water. Carrying what they could, the stunned passengers crowded to the ships gunwales, hoping they would survive long enough to reach the stretch of beach and wind-blown palm trees visible on the rocky shore. *(For more on the location of the Sea Venture, see Appendix B).*

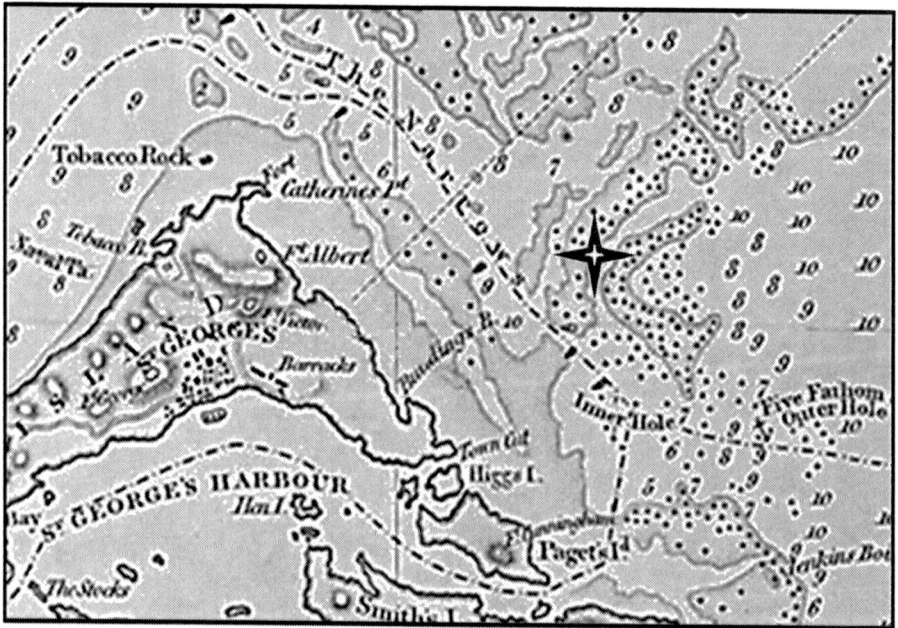

Approximate site of the Sea Venture wreck placed on the 1863 Sea Chart by R. H. Laurie; St. Catherine's Beach is at left center.

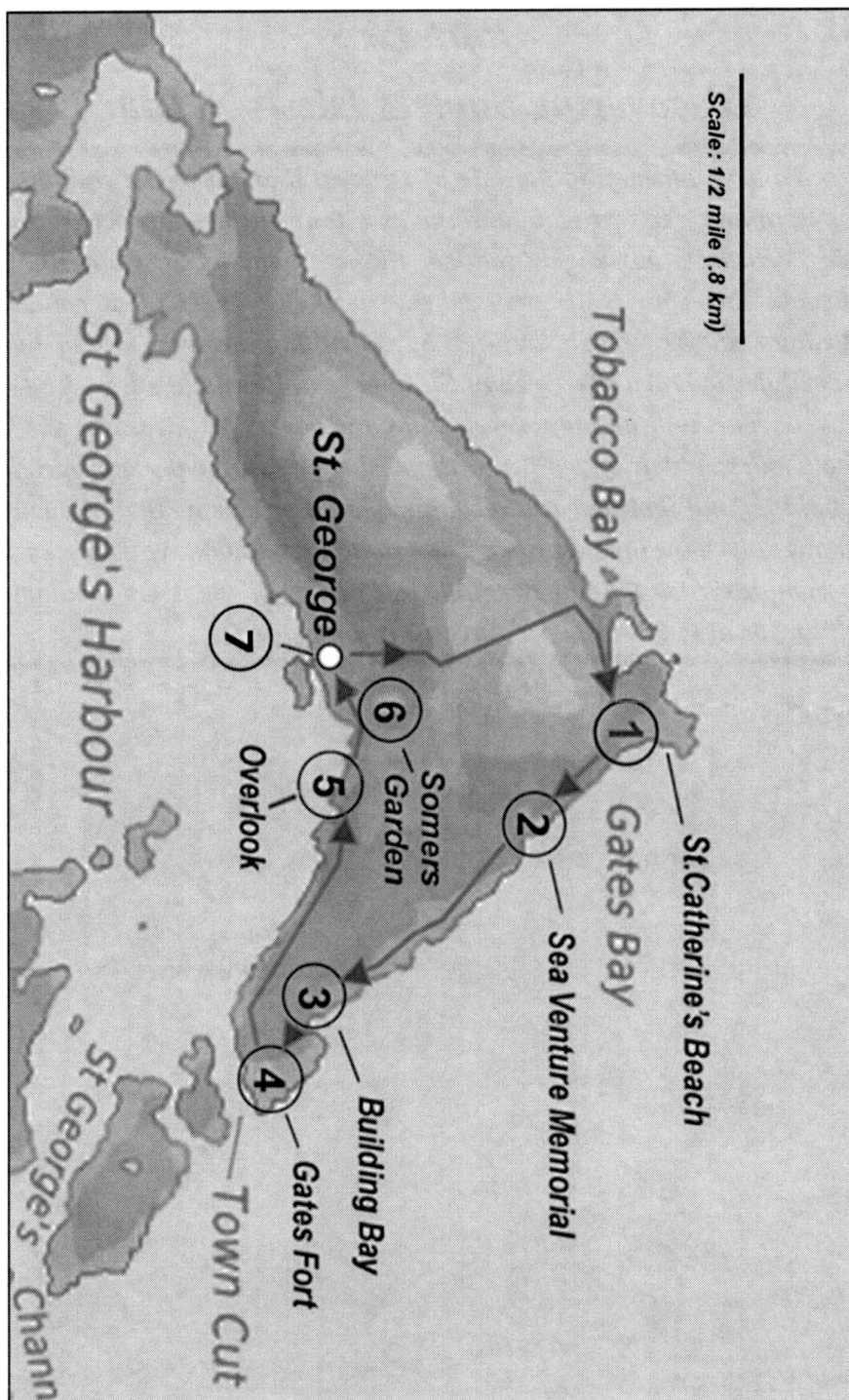

Scale: 1/2 mile (.8 km)

Tobacco Bay

St George's Harbour

St. George

St.Catherine's Beach

① Gates Bay

② — Sea Venture Memorial

Somers Garden

Overlook

⑦

⑥

⑤

③

④ — Gates Fort

— Building Bay

Town Cut

St George's Chann

Bermuda's East End – Overview of Tour Stops 1-7

Part Two

Discovering Somers Isles—A Tour

To best understand the tale of survival that led to the founding of Bermuda, the text will continue as a tour of the shore where the Sea Venture's passengers landed. Known today as St. Catherine's Beach, this spot is located on the northeast coast of Bermuda. Visitors can best reach the area by consulting the ferry and/or bus schedules to reach the Town of St. George in Bermuda's East End. *Note: This tour includes seven stops and covers just over two and a half miles; much of path is without shade: sun block, appropriate clothing, and footwear are recommended. For clarity, the following maps only show the described Tour route; in addition, you may wish a complete road map of Bermuda. (A variety of maps are available online or at the Bermuda Tourism Visitor Centers).

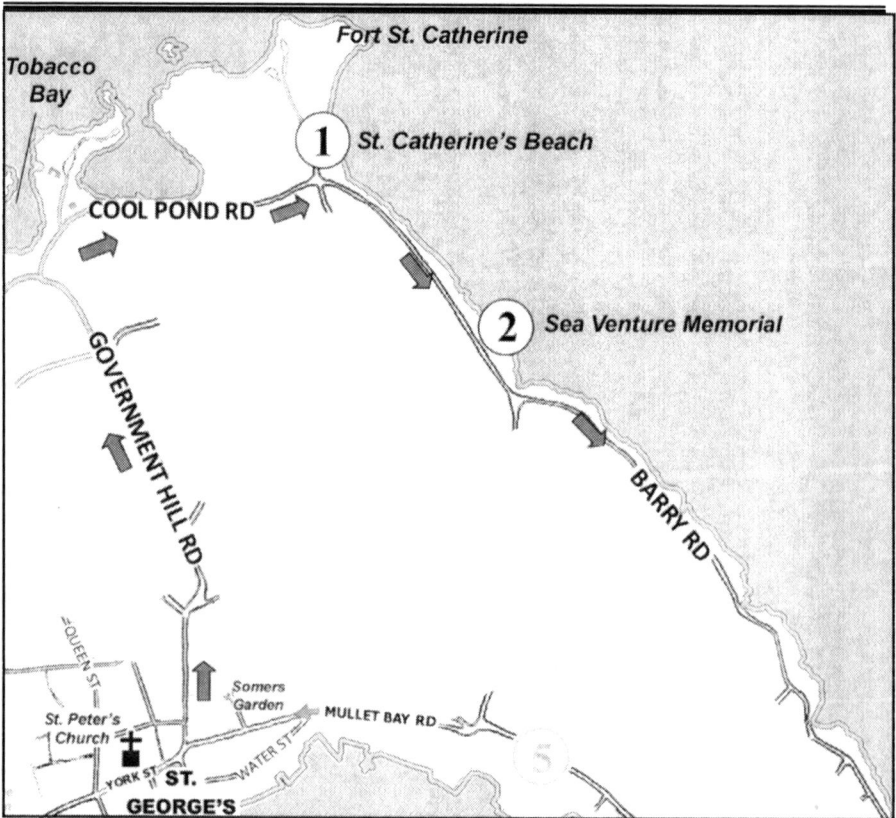

The town of St. George is a UNESCO World Heritage Site and you should plan to spend more time visiting here. However, the tour will return to the town later, and its early history discussed in context.

If you are traveling on foot, you may take a taxi or shuttle from King's Square in the center of St. George's to travel the first mile to the shoreline at St. Catherine's Beach. Or, if you wish to explore by bicycle or scooter, use the following directions:

- From King's Square turn right onto Duke of York (or York) Street, then turn left onto Duke of Kent Street / Government Hill Road.

- Follow Government Hill Road another 1/2 mile (.8 km) to the end above the beach at Tobacco Bay.

- Before the road descends to the beach, turn right on Cool Pond Road and follow 1/4 mi. (.4 km) to the intersection with Barry Road at St. Catherine's Beach.

STOP 1: ST. CATHERINE'S BEACH & GATES BAY

When you arrive at St. Catherine, face towards the water. About 1000 yards across the bay are the eastern reefs where the Sea Venture came to rest. To your left is Fort St. Catherine; although the site of an original 1612 fort, this defense has been rebuilt five times, and what stands today is the restored nineteenth-century British fortress. The museum here has numerous displays and artifacts, many related to this story, and is well worth a visit.

St Catherins forte
F

When you are ready, you may wish to find a comfortable spot at the beach to read the following.

Imagine yourself among the *Sea Venture's* passengers, exhausted, terrified, and seasick, waiting those long minutes aboard the stricken *Sea Venture* for your turn to take the ship's boat to shore. However, as a leader and one of the "better sort," it appears Thomas Gates made his way to the shore at the outset; some accounts describe him jumping into the surf there, shouting, "Gates—His bay!", which, in theory at least, staked his claim to the island for England.

Sir George Somers likely did not witness Gates' dramatic moment; if he had, he might have viewed it with some chagrin: right on the heels of the Admiral's exertions in saving the ship and all aboard, Gates was assuming command. Sir Thomas certainly had no doubts about his role: while at sea, Admiral Somers had been in command, but on land, whether on Virginia's shores or not, *Governor* Gates was in charge; the moment would mark a decline of the two men's relationship.

Before long, the tired crewmembers managed to ferry all the colonists to shore. Tossed at sea for four days, ill and dehydrated, no doubt many of the passengers collapsed, retching on the sandy shore. Somers and Captain Newport may well have been some of the last to leave the *Sea Venture*, for there was still much to do; the ship still stood relatively intact for the moment, and despite the water damage below decks, there were provisions that could be salvaged. With the storm receding and the passengers safely ashore, the officers now ordered the crew to begin to ferry unspoiled foodstuffs and valuable supplies to the beach.

By this time, the two veteran mariners had likely concurred: the only mid-ocean islands they could have reached were the Bermudas, the infamous "Isle of Devils." Given the already fragile condition of the settlers, for the moment, the officers may well have decided to keep this knowledge to themselves. 14

Stop 1: The view from St. Catherine's Beach today; the survivors from the Sea Venture gathered on the shore near this spot.

In 1609, the beach here was probably not as wide, but due to the lower sea levels of the seventeenth century, the waves washed the beach further offshore than they do today. When the colonists landed, a mix of scrub, Bermuda cedar, and palmetto trees filled the slopes behind you. What lay beyond them was an unknown, and surely unease began to spread amongst the exhausted survivors.

Amazingly, by the end of the day, all 150 survivors (including the ship's dog), and most of the supplies that could be saved were ashore. Perhaps it was some comfort for the shaken settlers when Reverend Buck gathered them together to offer a prayer of thanks for their salvation. However, at some point, word would begin to spread of where the *Sea Venture* had landed—and the islands' reputation. As Jourdain related, their refuge was, "the most dangerous, infortunate, and most forlorn place in the world." "A place so terrible to all that ever touched on them," wrote William Strachey, "such tempests, thunders, and other fearful objects are seen and heard about them, that they be called commonly, The Devil's Islands… feared and avoided of…any other place in the world." In retrospect, given these thoughts came from two educated men, we can only imagine the effect amongst the more superstitious colonists as the sun began to set. Whatever threat the darkening wood slopes above the beach held, as a precaution Governor Gates placed guards around the encampment for the night. Doubtless too exhausted to care what happened after nightfall, the colonists spent their first undisturbed sleep in five days on the shore of Bermuda.

As the morning brightened on Saturday, the *Sea Venture* was still visible, held fast by the reef, but relatively upright and stable. Somers ordered his crew to return to the ship again, this time to retrieve the remaining supplies and any salvageable parts. The odds on being found by the rest of their scattered convoy, much less a passing ship were small; if Gates and Somers were to continue to Virginia they would need to build a new vessel. To that end, spare sails and rigging, pitch, tar, the all-important ship's tools, even iron bolts and wood planks were stripped from the *Sea Venture*.

As the day brightened, Gates' military experience served him well in organizing an encampment on the shore. The Governor assigned the women and some of the men to build shelters far above the waterline. On the slopes above, they found materials ready at hand; in addition to the pervasive cedar trees mentioned by previous visitors, tall palmetto trees grew everywhere: a wild palm with broad leaves ideal for thatching a shelter's walls and roof. Other parties made their way through the thick foliage to explore the island for food and water. The results of their search must have stunned the survivors; accustomed to lives in urban England or at sea, what they found on Bermuda was not a demon's perdition, but a paradise. With scant human visitation and few natural predators, the amount of wild life was almost overwhelming. Not surprisingly, Admiral Somers looked first to the ocean for sustenance; there he "found such a-fishing, that in half an hour with a hook and line, he took so many as sufficed the whole company."

Originally intended to snare Virginia deer, nets salvaged from the

Bermuda's First Settlement, 1609 (Oil painting by Christopher M. Grimes)

ship were used to close off the mouth off the small inlet south of the camp, later known as Building Bay. There, as William Strachey described, they caught, "five thousand of small and great fish at one haul: as pilchards, breams, mullets, rockfish, etc., and other kinds for which we have no names." A few months later, tortoises would appear on the islands' shores, some large enough that, "one turtle feasted well a dozen messes, appointing six to every mess."

Nearby, other searchers discovered boar trails running through the woods; left on the islands years before by mariners such as Oviedo, the hog population had increased dramatically. In addition to the pigs that may have been aboard the *Sea Venture*, hundreds of boars were captured and collected in hastily built pens; fattened on palmetto berries, the pork would supply food on days the seas were too rough for fishing. Gates also directed that seawater be boiled in kettles from the wreck; the resulting salt would preserve meat for future use.

Birds were also plentiful and, unaccustomed to two-legged predators, unafraid of the new visitors. "Fowl there is great store," Strachey recounted, "small birds, sparrows fat and plump like a bunting...white and gray heronshaws, bitterns, teal, snipes, crows, hawks [and] cormorants." Fortunately, by that Fall the colonists had become more comfortable with their surroundings, for it was then that after sundown, the settlers heard the source of the island's evil reputation: the unearthly cries of the cahow. Far from demonic, they found the bird tame, almost submissive. Strachey continued:

A kind of web-footed fowl there is...which all the summer we saw not, and in the darkest nights of November and December ...they would come forth but not fly far from home and hovering in the air and over the sea, made a strange hollo and harsh howling.

Our men would take twenty dozen in two hours of the chiefest of them; and they were a good a well-relished fowl, fat and full as a partridge. In January we had a great store of their eggs, which are great as an hen's egg.

The Cahow (Bermuda Petrel)

Too docile for their own safety, the fowl were killed by the hundreds each week. (Considering the sheer number of birds captured, the egg consumption described in Strachey and Jourdain's accounts, and the fact that the cahow lays only one egg per season, it is no surprise within ten years these birds were almost extinct).

Fresh drinking water was to be more of a problem: the searchers discovered no streams or springs. But the islands' layers of sandy soil and coral had a side benefit; as one settler described, "For all or the most part of the fresh water cometh out of the Sea draining through the sand, or that substance called the Rock, leaving the salt behind, it becomes fresh." And, to collect water near the encampment, Gates directed that basins be dug on the rocky slopes to collect rainwater. William Strachey later recalled:

In August, September, and until the end of October, we had very hot and pleasant weather…many scattering showers of rain (which would pass swiftly over, and yet fall with such force and darkness for the time as if it would never be clear again) we wanted not any; and of rain more in summer than winter.

The storms left behind an ample supply of rainwater that proved invaluable in supplying their needs (as it does in Bermuda to this day). But other than the occasional storms Strachey described, the colonists found the Fall weather comfortable, and the following winter mild, particularly as compared to England.

Seeing the paradise from a different perspective, Sir George Somers began to investigate the islands as a potential resource for England. In the first weeks, he squared off a garden above the shore, and with some of the seeds intended for Virginia, he tested the fertility of the soil for lettuce, melons, peas, and eventually, sugar cane. During his experiments, Somers found the islands had few harmful insects or animals (except for the wild boar that apparently rooted up his first garden).

Using a flat-bottomed dinghy, Somers also began to explore, leaving for days at a time to map the island chain. As others had noted before him, Somers knew the Bermudas were not only an ideal way-station on the Indies' trade routes, but also found that only two harbors were accessible from the sea: the dreaded reefs surrounding the islands offered a natural fortification.

All in all, the resources were seemly endless, and the scenery unquestionably beautiful. After their perilous journey, the island's abundance, "caused many of [the colonists] utterly forget or desire ever to returne from thence, they lived in such plenty, peace and ease." Their leisurely days were about to end.

STOP 2: THE SEA VENTURE MEMORIAL

The tour will deviate briefly from the chronology of the settlement to visit the modern Sea Venture Memorial. When you are ready, turn right (south) from St. Catherine's Beach and continue down Barry Road about 1/4 mile (.35 km).

* CAUTION: There is no parking at this stop; please find a spot for your vehicle safely off the roadway.

Placed on the 400th anniversary of the 1609 wreck, the monument here is a replica of the "mnemosynon" erected just before the ships left Bermuda for Jamestown. Using timbers salvaged from the Sea Venture, the colonists fashioned a cross and using wooden screws, fastened it to a cedar tree. Attached to the cross were copper plates inscribed in Latin and English describing the Sea Venture's fate. The tablets on today's memorial also list the names of the known passengers aboard the Sea Venture (see Appendix). The placement of the original marker in 1610 will be discussed later in the tour.

If you wish, remain in this area to read the following text.

Within a month of the wreck of the *Sea Venture*, a number of changes were to occur that changed the complexion of the settlement. First, despite the beautiful scenery, moderate weather, and abundant food, Gates, Somers, and Newport, as well as those colonists who had large investments in the Virginia Company, had not forgotten it was still their duty to get to Virginia.

Knowing nothing about the fate of the rest of the convoy, Governor Gates was eager to advise the colony at Jamestown of their condition, and designate temporary leadership until he could arrive. The ship's longboat had already saved the lives of the settlers once, now it was to be re-fit with sail to make the journey to Virginia. Once there, mariners could then deliver Gates' messages, and arrangements be made for other vessels to return for the marooned settlers.

By the end of August, the *Sea Venture*'s longboat was decked over with hatches salvaged from the wreck, and with the addition of mast and sail, was felt to be seaworthy enough to survive the 600-mile journey to Jamestown Fort. Henry Ravens, the master's mate from the *Sea Venture,* would pilot the craft, along with a six-man crew. The Company's cape merchant, Thomas Whittingham, whose role had been to oversee Jamestown's stores and trading, was now entrusted with letters from Gates designating temporary leadership over the colony until he could arrive.

Ravens set sail with the pinnace on August 28th, pledging to return with a larger vessel in a month. However, Ravens was back two days later, unable to negotiate the islands' shoals to the northwest and east. With fresh supplies, Ravens' party left a second time, this time successfully leaving through the channel near where the *Sea Venture* ran aground; the group was never heard from again. (Ravens' and his crew likely died at the hands of the Powhatan within a few miles of Jamestown, but their fate would not be learned until after the colonists reached Virginia).

Whether the goal of sending Ravens in the longboat was merely to relay information—a "bark of aviso," as Strachey described it—or Gates was doubtful about the success of the desperate mission, he ordered a new vessel built the very day Ravens departed. An accomplished carpenter and shipwright, Robert Frobisher was to design and build a larger ship, using the sheltered cove just south of the encampment for the construction. Doubtless, Governor Gates also believed work on a common project would build unity, but it was not to be.

Undercurrents

By the end of August, a more perilous change was occurring in the settlement: undercurrents of dissention. The first came soon after reaching the island: a simple sailor's brawl between Robert Waters and Edward Samuel turned to murder when Waters took the blade of a shovel to his opponent's head. Gates had the man arrested, and without a proper jail, tied the prisoner to a tree with his victim for the night. Apparently more popular than his victim, Waters was released by comrades after dark, and the murderer escaped into the woods.

Soon after, Admiral Somers intervened on his crewman's behalf with Gates, and "upon many conditions," Waters was returned to camp.

The next challenge to Gates' authority occurred soon after. Considering the abundance of the islands, some in the settlement began to question why their leaders were making plans to leave the Bermudas. The work on a new ship led several of the crewmen, soldiers, and indentured workers ("the common sort," Strachey described), to believe resuming the voyage to the dreaded Jamestown was an unpleasant option. One crewmember, John Want, began convincing others to stop work on the pinnace; according to his plan, the group would leave those loyal to Gates behind and live in peace on another one of the islands.

On September 1st, the Governor discovered the mutiny. The principal mutineers, Want and five other mariners, were given their own island by Gates, but it was not one they would have chosen. The barren islet selected for them soon wore the rebels down. The Governor eventually forgave these six, but it wasn't the last problem Gates would have with "the common sort." 15

Once the keel was laid for the new vessel, soon to be known as *Deliverance*, the construction began using salvaged parts, as well as fresh-cut cedar. What Sir George thought of the project is not recorded, but while Gates, his soldiers, and settlers labored in the September sun, Somers had his crew continue fishing to feed the settlement, while he went on with his exploration and mapping of the islands. Ironically, at the same time, Governor Gates made plans to salvage the failing colony at James Fort, Sir George Somers was laying the foundation for a future colony at the Bermuda Isles.

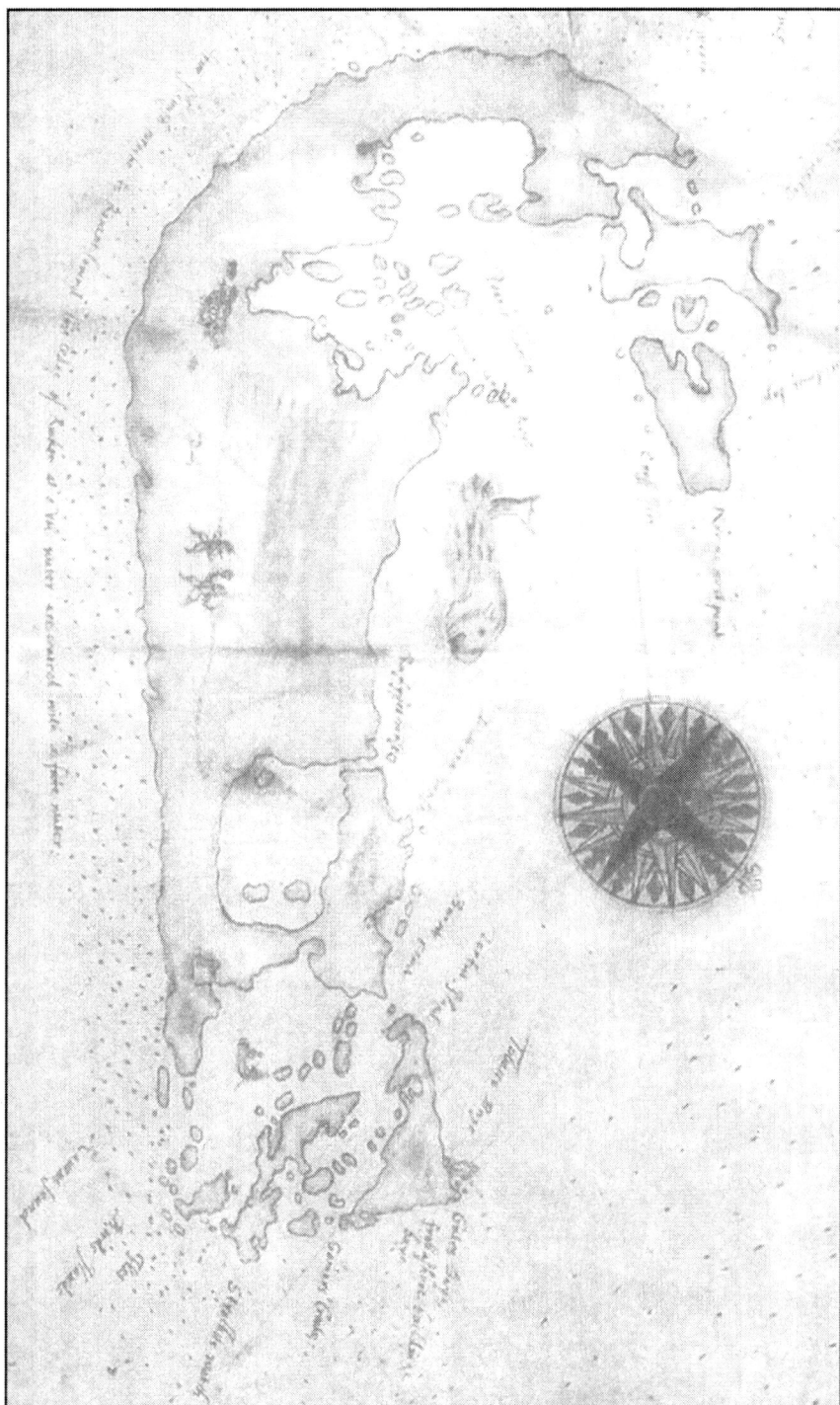

Bermuda as drawn by Sir George Somers, 1609 (Bermuda National Trust Collection)

STOP 3: BUILDING BAY

The tour will now visit the inlet where the new ship was to be built; known popularly as Alexandra Battery Beach, to this day the cove is still also known as Building (or Buildings) Bay. As the former name implies, the area was adapted as a fortification by the British military in the nineteenth century.

When you are ready, continue south down Barry Road for 6/10th mile (.9 km) to the next beach on your left. Parking is available for Building Bay at a pull-off on your left.

By the Fall of 1609, the cove in front of you would be a beehive of activity. Stocks at the shoreline held the keel, ribs, and planks that made up the growing hull of Gates' new vessel.

If you wish, stay in this area for the following text.

It would not take a sailor to realize that the new forty-foot keel represented a vessel not only half the size of the *Sea Venture,* but also less than half the carrying capacity. According to Strachey's account:

> *Sir Thomas's determination to move on to Jamestown caused an undercurrent of disagreement throughout his time in Bermuda. To make life a little more difficult, he had to try to dampen down a rumour that he intended to use the Deliverance to take only those close to him to Jamestown, leaving the rest behind.*

Perhaps as mariners, Somers and his men were more optimistic of the success of an experienced seadog like Henry Ravens; but as November of 1609 waned with no word, it became apparent their comrade had come to misfortune. On November 27th, Sir George proposed to the Governor that they start work on another vessel. For this project, obviously Somers would require manpower, tools, and

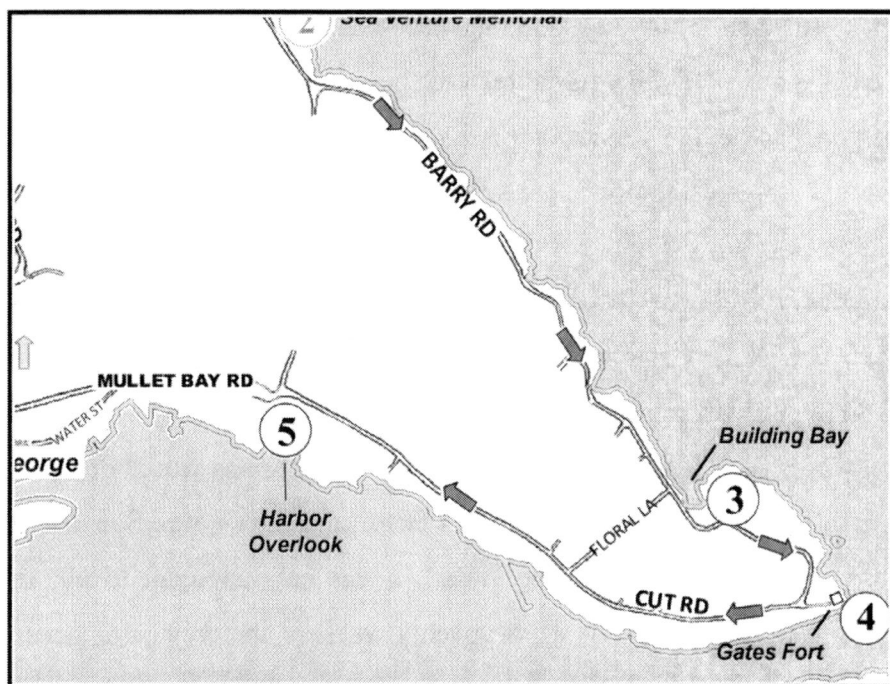

Bermuda's East End – Stops 3,4,5

Building Bay, known today as Alexandra Battery Beach (2007)

hopefully, some of the valuable pieces salvaged from his wrecked ship. Whether Sir Thomas agreed to the wisdom of the idea in spirit, in practice he apparently felt his pinnace was the more important. He allowed the Admiral two of the four carpenters, twenty men (in all probability, Somers' crew anyway), such tools as Gates' crew didn't need, and only one iron bolt; and, as the Governor's ship would require all the salvaged wood, the new vessel would also have to be built entirely from the island's cedar trees.

In the years that followed, a number of historians would describe a growing rift between Sir Thomas Gates and Sir George Somers. If one did not exist before this point, it was now apparent to all. In addition to the departure of the twenty workers and crewmen, surely Somers arranged for those close to him, and likely others who disagreed with Gates' leadership, to accompany his work party. The Admiral and his group packed their belongings, and shifted their camp to the "main island" to begin construction of the second vessel.16

Perhaps as a matter of protocol, or due to their often reserved writing style, neither Strachey nor Jourdain's otherwise detailed accounts dwell on this rift in the leadership. We can well imagine that the departure of Admiral Somers with almost a quarter of the settlement—including some of those most knowledgeable in fishing and survival skills—must have been a shock to the community. As one later account related:

Now although God still fed them with this abundance of plenty, yet such was the malice of envy or ambition, for all this good service done by Sommers, such a great difference fell amongst their Commanders, that they lived asunder in this distresse, rather as meere strangers then distressed friends; but necessity so commanded..."

The matter settled, Sir Thomas, his soldiers, and most of colonists remained at the original settlement. Life within the Governor's well-ordered camp went on: the duties of those who had left were reassigned, housing was rearranged, and the ship's bell still tolled morning and evening, and twice on Sundays to call the settlers to prayer. And woe to those who missed Reverend Bucke's sermons, which ironically included topics such as, "thankfulness and unity." William Strachey later recalled, "The names of our whole company were called by bill, and such as were wanting were duly punished."

Marooned, and with much to be done, the colonists still managed a personal life: just before the settlement split in November, marriage

Building the Deliverance, 1610
(Oil Painting by Christopher M. Grimes)

rites were performed for Somers' cook, Thomas Powell, and Mistress Horton's servant, Elizabeth Persons. In February: "the child of one John Rolfe christened, a daughter...and we named it Bermuda." Strachey added; "As also, the five-and-twentieth of March, the wife of one Edward Eason, being delivered the week before of a boy...and we named it Bermudas." (The Rolfe infant died on Bermuda; the Eason boy apparently survived).

At Building Bay, Sir Thomas continued to direct work on the new ship. In addition to adapting the *Sea Venture's* salvaged wood, cedar trees had to be felled, split, and long planks fashioned to form the hull—back-breaking labor, particularly for those inexperienced at such work. Not one to lead from a distance, Governor Gates himself stepped in. Even reading through Strachey's fawning account, it is plain Gates' leadership was an inspiration:

The governor dispensed with no travail of his body nor forbear any care or study of mind, persuading...an ill-qualified parcel of people by his own performance than by authority...to hold them at their work . . . His own presence and hand being set to every mean labor and employed so readily to every office, made our people at length more diligent and willing to be called ... And sure it was happy for us, who had now run this fortune and were fallen into the bottom of this misery, that we both had our governor with us, and one so solicitous and careful, whose both example (as I said) and authority could lay shame and command upon our people. Else, I am persuaded, we had most of us finished our days there.

In the meantime, Admiral Somers moved his party to the "main island" across the inner harbor (the exact location is not known); upon landing, Somers surely directed some of the group to build new shelters, while the rest were to start construction of the second vessel from scratch.

While several of Somers' men may have been accustomed to ship construction, the task was daunting. Using only the tools not required by Gates for the *Deliverance* construction, all the beams and planks for the new craft had to be shaped from scratch. Other than one large iron bolt salvaged from the *Sea Venture*, Somers' men had to fasten all the pieces with "trunnels," or tree-nails—wood dowels also shaped by hand.

Making the hull watertight was another challenge. Sylvester Jourdain described their ingenious solution:

> *The greatest defects we found there was tar and pitch for our ship and pinnace, instead whereof we were forced to make lime there of a hard kind of stone and use it...with some wax we found cast up by the sea from some shipwreck, served in turn to pay [seal] the seams of the pinnace Sir George Somers built, for which he neither pitch nor tar.*

However, as the two vessels began to take shape, amidst "an abundance of plenty," on one hand, and the growing probability of a renewed voyage to Jamestown on the other, dissention grew. This time, the dissatisfaction arose not from the "common sort" that had so concerned William Strachey. In the months previous, Stephen Hopkins' knowledge and ability to read scripture had gained him a

Modern view looking north from the Town Cut shows the relationship between Gates Bay (Stop 1, upper right), Building Bay/Alexandra Battery (Stop 3, center), and Gates Fort (Stop 4, lower left)

position as Reverend Bucke's clerk, and weekly Bible readings had made Hopkins a respected member of the settlement. However, in January, the clerk started preaching another belief: only God, not their leaders, had authority over the settlers. It is unknown how many colonists Hopkins swayed, but on January 24th, two of them, Samuel Sharpe and Humfrey Reede, reported Hopkins to the Governor. 17

Gates did not know the scope of the conspiracy, or if it had spread to Somers' distant camp; he promptly called the colonists together. As Strachey related: "[Hopkins] being only found, at this time, both the captain and the follower of this mutiny, and generally held worthy to satisfy the punishment of his offense with the sacrifice of his life, our governor passed the sentence of a martial court upon him." Not surprisingly, Hopkins pleaded for his life; his entreaties affected "the hearts of all the better sort of the company," including Captain Newport, and Strachey himself, who apparently now had some influence as Gates' acting secretary. Under pressure, the Governor again relented to the community, and Hopkins went free. William Strachey's ingratiating account of the pardon praised the Governor's wisdom because, "into what a mischief and misery had we been given up, had we not had a governor with his authority to have suppressed the same?" But the challenges to Gates' authority were not finished.

As the winter wore on, work continued on the ships despite storms that interrupted the otherwise temperate weather. Strachey later wrote: "In the beginning of December we had great store of hail (the sharp winds blowing northerly), but it continued not... Yet the three winter months, December, January, and February, the winds kept in those cold corners, and indeed then it was heavy and melancholy being there."

Perhaps that "melancholy" led to another a more widespread and dangerous conspiracy: members of both camps plotted to take over the settlement, kill Gates, and any who resisted their plot. Once again, some of the conspirators with cold feet revealed the plan to the Governor; as it happened, the mutiny was unmasked prematurely on March 13th, when Henry Paine, one of the "better sort" in Gates' camp, very publicly refused to perform guard duty, then struck an officer, and denounced Gates. Prepared by this time, the Governor's response to the defiance was swift. The mutineer was seized, brought before the camp, and the charges read; the sentence: execution. This time there would be no pardons; Gates allowed only Paine's request that as a gentleman, he be shot instead of hung. As William Strachey related, "toward the evening he had his desire, the sun and his life setting together."

Across the bay in Somers' camp, the other conspirators ("most of the camp," according to Strachey), assumed they had been given up by Paine and fled into the woods. Soon, Governor Gates received a message from the mutineers, this time asking that the group not only to be left alone, but also supplies be provided for them when the new vessels departed. Whether Gates believed Somers was involved, or merely the best suited to negotiate with his crew, he sent his response to the Admiral. Gates claimed that had they been stranded by the "tyrannie of necessitie"; in fact, the Governor had no intention of leaving them, "like savages," but intended to have boats return to rescue them (had this been his intention, Gates might have avoided some of the ill-will had he explained this from the outset). Addressing

Somers himself, Gates reminded the Admiral of their long friendship, as well as his duty to the Virginia Company to complete their journey. Eventually, at Somers' urging most of the men relented and returned to camp. Two mutineers already twice disloyal to Gates, Christopher Carter and Robert Waters, "by no means would any more come amongst Sir George's men...from which time they grew so cautelous and wary for their own ill." The pair would remain on the island when all the other colonists left.

Patience and Deliverance

Considering the events of the previous ninety days, it was with no little relief to Governor Gates that the construction of the *Deliverance* was almost complete. All that remained was to bring her to calm waters to season the hull, load ballast, and fit the rigging. As William Strachey recalled:

The thirtieth of March, being Friday, we towed her out in the morning spring tide from the wharf where she was built, buoying her with four casks...we launched her unrigged to carry her to a little round island lying west-northwest and close aboard to the back side of our island, both nearer the ponds and wells of some fresh water, as also from thence to make our way to the sea the better, the channel being then sufficient and deep enough to lead her forth when her masts, sails, and all her trim should be about her.

As the workers wrestled the Deliverance from the stocks that held her during construction, imagine their satisfaction when the new vessel slid into the bay here, ready to be completed in the deeper waters of the harbor to the west.

When you are ready, leave Building Bay and continue south down Barry Road; although there was no formal path here in 1610, perhaps some of the colonists followed the shoreline just as you are, watching the Deliverance as it was towed to the harbor.

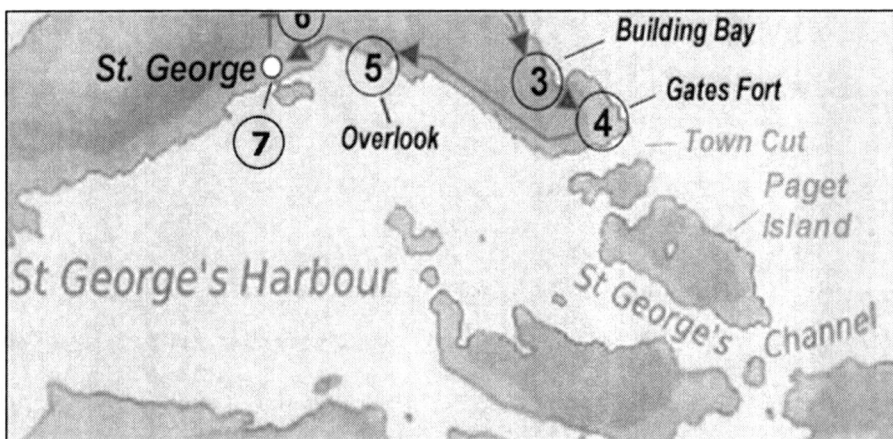

STOP 4: GATES FORT & TOWN CUT

In 250 yards (.21 km), you will reach Cut Road and the parking lot for Gates Fort. Park your vehicle here and continue up the sidewalk for a spectacular view of the Town Cut and the entrance to St. George's Harbor.

Similar to Fort St. Catherine, over time the fort here has seen a number of modifications. Although not built by Thomas Gates, the next Governor, Richard Moore, built defenses here between 1612 and 1615. Originally known as Davers (or Danvers) Fort, the post was rebuilt and named Gates Fort.

Below the Fort, the Town Cut has been widened over the years to accommodate today's larger ships. However, in 1609 the passage here was considerably shallower (on his map, Somers labeled it only as "Somers Creake"). To reach St. George's harbor to your right (west), even the diminutive Deliverance was probably towed further south past Paget Island and entered through what is now known as St. George's Channel.

STOP 5: ST. GEORGE'S HARBOR - OVERLOOK

Return to your vehicle and continue west down Barry Road for 6/10 mile (.9 km) to the intersection with Mullet Bay Road. Parking is available at a pull-off on your left, and allows you a panoramic view of St. George's Harbor. The town of St. George lies to your right (west); just off shore of the town stand the Customs buildings that now occupy Ordnance Island. In 1610, the Deliverance was towed into the harbor here, and possibly moored at one of the islets that were later combined to form the modern island.

You may remain in this area for the following text.

While weather improved during that April, work continued on the *Deliverance*. As Somers had anticipated, the new vessel was less than a third the size of the *Sea Venture*. William Strachey later wrote:

[The Deliverance] was forty foot by the keel and nineteen foot broad at the beam...she was eight foot deep under her beam; between her decks she was four foot and an half... She had a fall of eighteen inches aft to make her steerage and her great cabin the more large; her steerage was five foot long and six foot high, with a close gallery right aft, with a window on each side and two right aft. The most part of her timber was cedar...her beams were all oak of our ruined ship, and some planks in her bow of oak, and the rest as is aforesaid [cedar]. When she began to swim (upon her launching) our governor called her the "Deliverance," and she might be some eighty tons of burden.

Although there is no record of the location of the "little round island," by Strachey's description, it seems likely *Deliverance* was moored to an islet on the shoreline of what is now St. George's to be "nearer the ponds and wells of some fresh water." 18

While the *Deliverance* was rigged with sail and neared completion, the cedar pinnace built in Somers' nearby camp was also readied for launch. By the end of April, the appropriately named *Patience* sailed into the channel and Admiral Somers and his crew rejoined the other colonists. As Strachey recalled:

> *By the keel [she was] nine-and-twenty foot, at the beam fifteen foot and an half, at the luff fourteen, at the transom nine; and she was eight foot deep and drew six foot water, and [Sir George} called her the 'Patience.*

In that Age of Sail, until the easterly Spring breezes turned, the colonists' departure would be delayed several days. While the travelers awaited more favorable winds, the *Deliverance* and the *Patience* were loaded with water drawn from the hillside wells, salted meat and fish prepared at Gates' camp, and the remaining supplies necessary for the last leg of the journey.

In the meantime, Governor Gates had a memorial erected overlooking their first camp and landing site. As described earlier (*see* ***STOP 2***), a wooden cross was mounted on a cedar tree that stood where Somers planted his garden in the first weeks on the island; a silver coin bearing King James' image was placed in the middle of the cross, symbolically laying claim to the islands for England. For all to read, a copper plate inscribed in Latin and English was fastened to the cross:

Looking west toward St. George's; Ordnance Island is at center

*The 1830's appearance of the Town of St. George was painted near
the above overlook (From original painting by Thomas Driver)*

In memory of our great deliverance, both from a mighty storm and leak, we have set up this to the honor of God. It is the spoil of an English ship (of three hundred ton) called the "Sea Venture," bound with seven ships more (from which the storm divided us) to Virginia, or Nova Britannia, in America. In it were two knights, Sir Thomas Gates, Knight, governor of the English forces and colony there, and Sir George Somers, Knight, admiral of the seas. Her captain was Christopher Newport; passengers and mariners she had beside (which came all safe to land) one hundred and fifty. We were forced to run her ashore (by reason of her leak) under a point that bore southeast from the northern point of the island, which we discovered first the eight-and-twentieth of July, 1609.

By May 10th, the easterly winds finally relented. At the outset Admiral Somers and Captain Newport led the way in the ship's boat, buoying the channel to the east, and the two small ships started to make their way into open water. The channel was narrow, "no broader from shoals on the one side and rocks on the other than about three times the length of our pinnace [about 40 yards]," and did not allow much room for error; we can well

imagine their horror when the winds failed them, and the *Deliverance* drifted, striking the rocks on the channel's south side. Fortunately, the sturdy ship took the blow and after scraping over the crushed rock, continued out to sea.

As the May breezes filled the sails of the two small vessels, they passed slowly across the harbor in front of you. Beyond Paget Island to your left front is St. George's Channel, and the colonists' path to the open seas beyond.

Aboard the *Patience* and *Deliverance,* even the most dedicated of the 140 colonists must have gazed with some regret at the islands that had been their salvation for ten months, for ahead lay Jamestown with its rumors of famine and death. Sir George Somers must have looked on with mixed feelings as well: the islands he had explored so carefully, a new colony perhaps, were fading in the distance. He would soon see them again.

As the other colonists sailed away, Christopher Carter and Robert Waters probably made their way to the high ground above the original camp. Watching the ships disappear, no doubt the mutineers wondered when, or if, they would ever return to England; their choice to remain behind rather than face punishment left the possibility of a lifetime marooned on the islands. By the late Spring of 1610, for all practical purposes, the Bermudas had returned to what they had been so many times before: a desolate archipelago in the middle of the Atlantic sheltering stranded mariners.

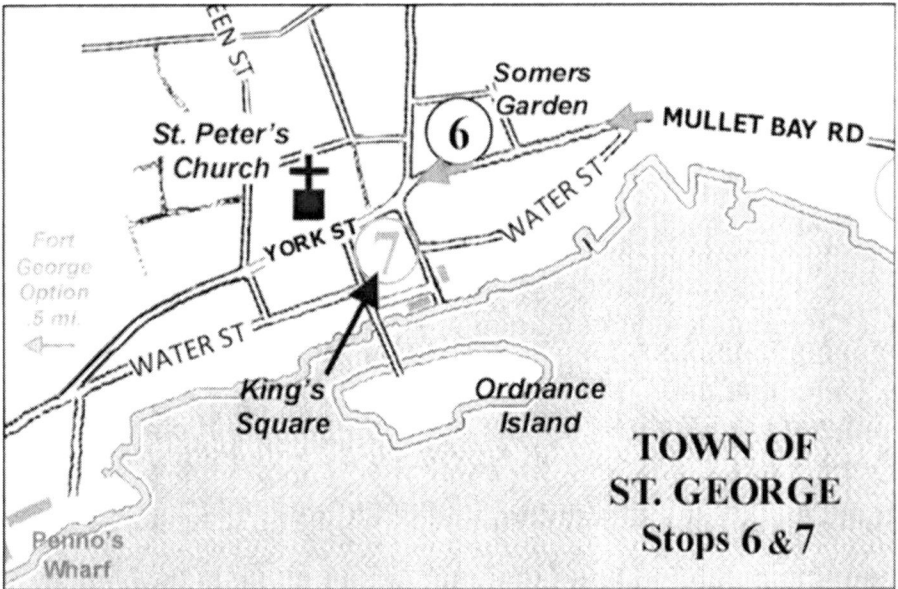

TOWN OF
ST. GEORGE
Stops 6 & 7

STOP 6: SOMERS GARDEN

When you are ready, continue west on Mullet Bay Road to the Town

of St. George. In approximately 1/4 mile (.35 km) the road turns into Duke of York Street; continue until you see an obelisk and walled park on your right. Park your vehicle in a space on the left and carefully cross the road to enter Somers Garden. The large monument at the entrance honors Sir George Somers and the 300th Anniversary of Bermuda's settlement.

To the left of the entrance steps below the obelisk, note there is another memorial, this one erected in 1876 by

Governor J.H. Lefroy. As described in the following text, tradition places it near where Admiral Somers' heart was said to be buried.

If you wish, take some time to explore the Gardens and then find a comfortable spot to read the following section.

On to Jamestown

Over the next week, the two vessels saw favorable winds, and at midnight of May 20th, "we had a marvelous sweet smell from the shore...strong and pleasant, which did not a little glad us." Soundings of the water indicated a shallow and sandy bottom, and at dawn, the cry of "Land" came from the lookout in the ship's foretop.

Upon reaching Point Comfort at the entrance to the James River, the new colonists were greeted with the happy news that the rest of the original fleet had weathered the hurricane, and successfully completed their voyage to Jamestown. "However," continued Strachey, "our governor had new, unexpected, uncomfortable and heavy news of a worse condition of our people above at Jamestown." While the *Sea Venture's* survivors had lived in abundance on Bermuda the previous winter, of all the previous Jamestown colonists—some 450 souls—only 60 had survived the "The Starving Time." And, despite the supplies brought from Bermuda, it was clear to Gates that the colony could not survive much longer on its own.

With the decision made to leave the settlement, only the chance arrival of a new supply convoy under Governor Thomas West, the Lord De La Warr, kept Jamestown from being abandoned. Even so, it was clear surviving through the coming winter was still in question; as the Bermudas were a certain source of foodstuffs, Admiral Somers offered to sail back aboard the *Patience*. In a letter dated June 20th,

1610 to the Virginia Company, Somers placed an optimistic spin to the moment:

> *Now we are in a good hope to plant & abide here for here is a good course taken & a greater care than ever there was. I am going to the Bermuda for fish & hogs with 2 small Pinnaces & am in a good opinion to be back again before the Indians do gather their harvest. The Bermuda is the most plentiful place that ever I came to, for fish, Hogs and fowl.*

Accompanied by a second vessel under Captain Samuel Argall, Somers left Jamestown on June 20th (or 19th, according to Strachey). Their voyage encountered unfavorable weather, such that Somers changed their destination to head north to upper "Virginia," in this case, the fish-laden shoals off Cape Cod. The two small ships struggled to keep together, but eventually reached the Sagadahoc

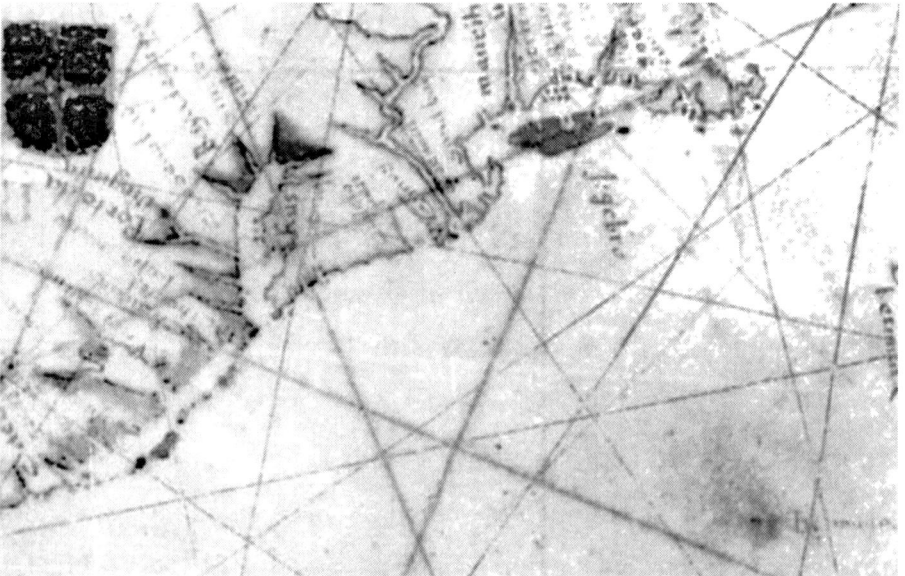

Section of Pascoal Ruiz' 1633 Portolan Chart of the Atlantic shows Virginia at middle left and Bermuda at lower right (LC)

(now Kennebec) River in Maine.

With the *Patience* loaded with fish, on July 26th Admiral Somers informed Argall he was ready to set sail again. Turning south, they again ran into poor weather, and the ships lost contact in the fog. Captain Argall finally reached Jamestown with his catch at the end of August, but what had happened to Admiral Somers and the *Patience* was doubtless a new source of concern. [19]

At some point after losing contact with Argall off the shores of New England, Sir George Somers decided to reattempt his voyage to Bermuda. No one recorded the date that Somers small craft reached the islands, but given the Atlantic's unpredictable summer weather, and the small amount of sail available on the *Patience*, it is unlikely he could have returned before early September. Given Argall's account of their somewhat odd voyage—not to mention Somers' mysterious behavior—speculation arose then, and since, that the Admiral had his own reasons for returning to Bermuda. [20]

Return to the Bermudas

On reaching the islands, Somers likely directed the *Patience* to the shelter of St. George's Harbor. He soon discovered the castaways, Christopher Carter and Robert Waters, not only healthy but thriving. Unlike, Somers' previous attempts, the pair had successfully planted crops in the islands' quirky soil. However, Sir George's exertions for the Virginia Company in Bermuda and at Jamestown caught up with him; on November 9th, Sir George Somers died at age fifty-six.

Regardless of his own intentions for Bermuda, when Somers fell ill, he requested that his nephew Matthew captain the *Patience* back to Jamestown with a load of supplies. Soon after his uncle's death, Matthew Somers indeed reloaded the cedar pinnace, but instead headed for England. The Admiral's remains were embalmed, and his body brought home to be buried with honors at Lyme Regis. Also aboard was Robert Waters, the twice disloyal mariner who had remained on the islands with Carter. Apparently fed up with a settler's life, even on Bermuda, Waters was willing to take his chances returning to England.

Christopher Carter, however, decided to remain behind; along with Somers' former aide, Lt. Edward Waters, and another of Somers' crew, Edward Chard, the three were to maintain England's claim to the islands. A poignant tradition holds that in keeping with his last wishes, Somers' heart was also left on the island, buried on the hillside where the town of "St. George" was soon established. 21

The Somers Isles

Ironically, at about the same time as Sir George Somers returned to Bermuda in September of 1610, Governor Gates was returning to England. With Governor De La Warr firmly in charge, Gates and Captain Newport returned to London with good news for the Company's investors: they and the other *Sea Venture* passengers had survived the hurricane, and the colony at Jamestown, if shaky, was back on its feet; but it was the news about the Bermudas that grabbed the attention of Company investors. The islands reportedly showed great potential for a wide variety of crops, wood, pearls, and whale oil—in sum, what Jamestown lacked: profitability. The written accounts brought back to London—Sylvester Jourdain's well-written prose, and a poem of survival penned by passenger Robert Rich—were quickly published, making it clear that the *Sea Venture*'s chance landing on the "*Devil's* Isles" was, to the contrary, nothing short of *God's* will.

But another account, perhaps not as glowing as the other two, also

"*Near this spot was interred in the year 1610, the heart...of Sir George Somers. Today's marker was placed on the site in 1878.*

made the rounds: William Strachey's lengthy letter to one of the investors. Unlike the other two portrayals, his richly detailed description also included an unhappy tale of mutiny and murder, grim details that made the descriptions of the island's wealth all the more believable.

In response, on March 12, 1612, King James granted the Virginia Company a Third Charter. In forty-nine pages of seventeenth-century legal jargon, the charter reaffirmed the Company's claim, and gave more power to local authorities to deal with mutinous behavior. Heavily in debt, the Company was also authorized to hold a lottery to raise funds; but most importantly, the charter extended the potential borders of "Virginia" another three hundred leagues east—about 900 miles into the Atlantic—to include the Company's claim to what were now officially known as "Somers Islands."

The Company modified their expectations for the colony as well: unlike Jamestown, no gold or vast wealth was anticipated (at least initially)—the settlement, and its profits, were to be primarily agricultural. Each settler was to be allowed about a quarter of an acre to farm (married men were given half an acre); in return, the colonists were indentured for seven years of work for the Company. To oversee their investment, the Company officially designated Richard Moore as Governor of Bermuda. Moore was a good choice; no wealthy gentleman of title, or Company speculator, Moore was previously a ship's carpenter, pragmatic, and "an able and resolute man."

Almost three years after the *Sea Venture* ran aground off the East End, on July 11, 1612, the English ship *Plough* sailed into St. George's Channel, anchoring in the small bay south of Smith's Island.

Aboard was Moore, the first designated English governor, and the first *intentional* settlers for the "Plantation" of Somers Islands. Arriving in Smith's Bay, about fifty new settlers stepped onto the beach and, as the colonists had in 1609, offered a prayer of thanks; only on this occasion, local inhabitants greeted them: Carter, Chard, and Waters. Scarcely clothed, but apparently in good shape, Governor Moore found them, "civil, honest, and religious, and making conscience of their ways ...they have planted corn, great store of wheat, Beanes, Tobacco, and melons, with many other good things for the use of man." The trio was industrious, yes, but after being marooned for two years, the men were also preparing to leave the islands, having cut and split cedar to build a ship to return to England.

Not the least of their motivation to get home was their discovery of 180 pounds of ambergris, the valuable excretion of sperm whales used in perfume. In London, the substance was worth some three pounds an ounce—but on Somers Islands, it was also Company property. Moore soon discovered the trio's plot to conceal the prize, and eventually imprisoned Chard as the ringleader. However, Governor Moore eventually used the ambergris for his own ends: by sending only a piece at a time to London, he not only showed a profit for the colony, but also guaranteed another supply ship would return for more ambergris.

After remaining a few weeks on Smith's Island, Moore moved the settlement across the harbor to the shoreline where the *Deliverance* had been completed two years before. Although crops thrived on the smaller island, Moore realized the area didn't promise the potential or defensibility of the larger island to the north. By August of 1612, the first shacks of cedar and thatched palmetto leaves appeared in the small valley above the harbor. Soon conveniently named for England's patron saint, as well as the islands' first benefactor, Sir George Somers, "their prime towne" was to become the oldest continuously inhabited English town in the Eastern Hemisphere.

Entering King's Square at York Street, St. George's

STOP 7- THE TOWN OF ST. GEORGE

The following locations are all within the Town of St. George and comprise Stop 7.

KING'S SQUARE:

When you are ready, return to your vehicle and continue down Duke of York Street for another 100 yards (.1 km), then turn left to return to King's Square. For your convenience, the Visitor Information Center and restrooms are located to the far left (east) side of the square.

Your tour will return to King's Square shortly, but you may wish to take some time from your tour to relax and explore this historic area. (At the end of the tour is a listing of nearby museums and sites). Alongside today's tourist attractions, several of the structures here date to the 1700's; several of these buildings doubtless owe their longevity to their Bermuda limestone construction. However, in 1612 this shoreline and the hillside above would be dotted with small shelters constructed of cedar and palmetto.

Find a comfortable spot in this area to read the following section.

Under Richard Moore's three-year commission as governor, the settlement on Somers Islands grew rapidly from a camp to a colony, albeit not without difficulties. The Governor had the settlers construct homes in St. George's of cedar saplings and palmetto leaves—a practical, though fragile, architectural format that would not change dramatically for years. In addition to shelter for the settlers, Moore ordered storehouses to be constructed (along with new corn and potato crops to fill them), and on nearby Water Street, the first Government House for administration was erected.

In accordance with Company wishes, the rest of Somers Islands was to be explored as well. In June of 1613, and again in 1615, the investors sent a "Master Bartlett" to start a survey with the intent of dividing the island into shares. However, finding that Bartlett had no instructions to include shares for those already settled on the island, Governor Moore sent the surveyor back to England on both occasions. However, the Company had also dispatched a multi-talented, 23-year old named Richard Norwood to the colony. As the inventor of a diving bell, Norwood's job was to search for the wealth in pearls that the investors still believed lay just offshore. When this project came to nothing, it was Norwood that was tapped by Governor Moore to perform his own survey of the islands.

According to the plan, St. George's and the eastern islands were to be maintained as common property, while the balance of the island were to be divided into eight equally sized "tribes." Known today as parishes, each tribe was named for one of the principal Company adventurers: James Hamilton, second Marquis of Hamilton; Sir Thomas Smith; William Cavendish, first Earl of Devonshire; William Herbert, third Earl of Pembroke; William Paget, fourth Lord Paget; Robert Rich, second Earl of Warwick; Henry Wriothesley, third Earl of Southampton; and Sir Edwin Sandys. Norwood's subsequent work was to have as much (or more) of an impact on the colony as any Briton to step onto the Somers Isles. 22

St. Peter's Church; although rebuilt several times, parts of the church date to 1620

The Virginia Company was specific in their other instructions to the Governor as well. First, when the settlement was established, it was expected that Moore would maintain a "religious government" of the colony. Similar to the 1609 settlement, prayers were to be read morning and evening, with those in attendance noted. As a meeting place for the Sabbath, a church of wood was built on the hillside above the camp; however, this structure was destroyed by a storm later that year, and Moore had another built of palmetto closer to the town. What came to be known as St. Peter's Church was later rebuilt on the same site; completed by 1620, it became the oldest Protestant church in continuous use in the New World.

Ironically, Governor Moore's "religious government" was to have similar disagreements with the clergy as Thomas Gates experienced in 1610. The new minister for the colony, George Keith, began to appeal to the malcontents, preaching that Moore, "did grinde the faces of the poore," with "Pharoah's taxes." The Governor promptly called the settlement together, and when it became apparent to all that Keith's viewpoint was not popular, the minister ended up much as Reverend Hopkins had in 1610: begging forgiveness.

SAINT PETER'S CHURCH:
When you are ready, walk from the Square back to York Street, this time turning left. Diagonally across the street, you will see a broad flight of stairs leading to St. Peter's Church. Although the church was rebuilt in 1713 due to damage from yet another storm, the earliest parts of the altar in this beautiful structure reportedly date back to 1620.

As mentioned, St. Peter's is still an active church, however, if not in use for services, take a few minutes to visit (entrance is free

St. Peter's Cemetery

"Belfry Tree" Ancient cedar in St. Peter's Cemetery

but donations are appreciated). Inside, you will find the solemn memorial plaques and unique cedar construction speak volumes about the Church's role in Bermuda's history.

In the churchyard behind St. Peter's is one of Bermuda's oldest cemeteries. Although the early graves are either unmarked, or their inscriptions long since worn away, doubtless some of the burials here date to the 1600's. With few exceptions, St. Peter's cemetery has been closed to further interments since the 1850's.

As you enter the cemetery, notice the wall to your left: segregated in death as they were in life, this section is dedicated to burials of the islands' slave and free black population. Despite their age, the inscriptions on many of these stones can still be read; the sentiments are often poignant: "...He was well esteemed by all who knew him."

Beside the walkway that winds behind the church, note the fallen cedar tree. Toppled during Hurricane Fabian in 2003, tradition has it this tree held the church's bell before the belfry was added in 1766.

Find a comfortable spot in this area to read the following section.

If Richard Moore had a failing, it was in following Company instructions too well. For next in Moore's commission—and, unfortunately, *before* instructions on the planting of crops—the Company required that, "especiall care must be to...sett yourself to fortifyinge" the islands. Underlining the message, at year's end another Company vessel brought a warning for Moore to prepare, "with all expedition," for it was believed the Spanish, "ere long would visit them." Alone in the middle of the Atlantic, and with little ordnance, "fortifyinge" is what Moore certainly did—almost to the exclusion of all else. 23

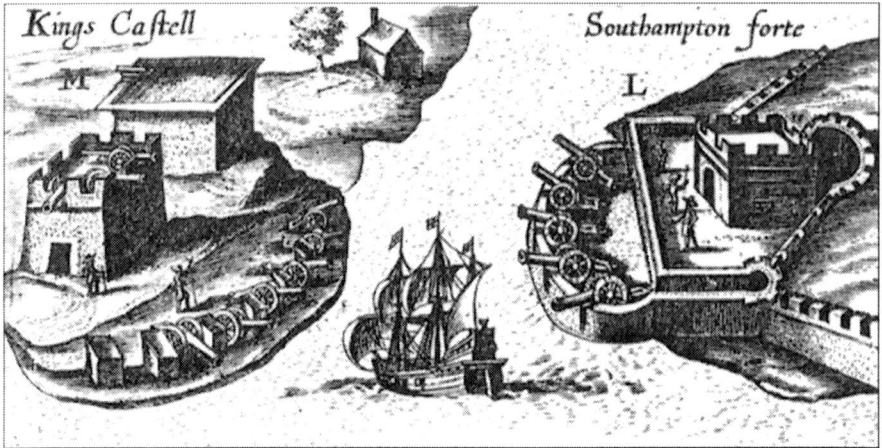

Initially only cedar-walled redoubts, small forts sprang up on Paget and Smith's Islands guarding the approaches to the St. George's Channel; to the south, overlooking the more recently discovered deep-water entrance of Southampton Harbor, King's Castle and Southampton Fort were erected. Overlooking St. George's itself stood "Riche's Mount," a watchtower where warning could be given of any vessel on the horizon. (*See Optional Stop for Fort George*). In all, the foundations for at least eight forts were in place by the end of 1613.

All their effort was not in vain: as apathetic as Philip III of Spain had been about the Bermudas, he finally began to be curious about the English settlement. In March of 1614, two Spanish ships attempted to enter Castle Harbor; there they were met by a pair of cannonballs from Moore's new forts. Little did it matter that the English gunners had but one shot left, and in the excitement spilled their only keg of gunpowder: the Spaniards had seen enough of the island's defenses and withdrew. Although not realized as such at the time, the event marked Spain's last attempt to retrieve the Bermudas from England.[24]

In the distance above St.George's is Fort Hill and the Maritime Operations Center; Riches Mount stood on this hilltop in 1613.

OPTIONAL STOP: FORT GEORGE

The site of the watchtower known as Riche's Mount is located atop Fort Hill about 1/2 mile from St. Peter's Church and may be visited as an optional stop.

* (If you wish to remain in the area of the Church, please skip the below directions and remain nearby to read the section that follows).

To visit the Fort (and its spectacular view of the harbor), return to York Street and turn right (i.e., away from the Square), and proceed .25 mile to where Fort Hill Road bears to the right. Follow Fort Hill Road around to the top. USE CAUTION: Fort Hill Road is narrow and steep; be aware of pedestrians and turning traffic.

Similar to Fort St. Catherine and Gates Fort, Fort George has changed many times over the years. After the tower was destroyed during a 1619 hurricane, the fort was rebuilt several times up through the 1700's before today's parapets were built about 1840.

Today, the fort still serves Bermuda as the Maritime Operations Center providing a variety of support for the islands and vessels in the surrounding waters.

Note: The other defenses started under Governor Moore are in locations beyond the scope of this tour; more information about these defenses may be found online, or by visiting the Bermuda National Museum in the Royal Naval Dockyard.

Remain in this area (or by St. Peter's church) for the following text.

All considered, for almost a year the colony on Somers Islands thrived. Under Moore's direction, a structured settlement of dwellings, a church, and storehouses stood in the shelter of St. George's Harbor; a local government was in place, and substantial work completed on the forts, and crops of corn, potatoes, and tobacco.

However, despite its early promise, the colony became beset with problems, ironically, many created by the plans of the distant London Company. Notwithstanding the early "victory" over Spain, Company instructions underscoring the importance of defenses diverted workers from the planting of new crops necessary for the coming winter. With each supply ship that entered the harbor, new settlers arrived from England, adding more mouths to feed; similar to others recruited by the Company, many of the new arrivals were either "gentlemen," or the "common sort" from London's slums, and unaccustomed to planting crops, much less building and maintaining a colony. 25

Not surprisingly, despite the relative success of crops of corn and potatoes (and inedible tobacco), two years of poor weather and the lean harvests meant the storehouses were starting to empty, and the following winters both saw famine. To minimize the demand on their limited food supply, Moore dispersed the settlement, sending one group to fish at Somerset, others to the southern islands, to feed on an already diminished cahow population. The tactic apparently succeeded; although some settlers certainly perished, when Minister Lewis Hughes arrived in 1614, he noted that the original sixty passengers from the *Plough* were still alive, an extraordinary feat of tenacity when one considers the hundreds who died in the famines at Jamestown, and perhaps testament to the islands' recuperative benefits.

However, to make matters worse, another competitor for the islands' limited food supply appeared: wood rats. Whether surviving a shipwreck years before, or arriving on one of the more recent vessels, the rat population began to overrun the settlement; the crops and stores of grain that Moore had been able to lay in soon fell prey. By June of 1615, well over 500 more colonists—and thousands of rats—vied for the islands' limited food supply.

Initially, Moore's request for supplies for Somers Isles fell on deaf ears. In withholding the ambergris and preventing the first survey, Moore had done himself no favors in London. All was well when those pieces of ambergris were generating profits (perhaps $2 million in today's money); but with little gauge of success other than profits, the Company now saw the reports from *another* struggling colony as Richard Moore's failing. As a contemporary historian recorded, "Now all the worst could possibly be suggested, was too good for him; yet not knowing for the present how to send a better, [the Company] let him continue still, though his time was near expired."

Replica of the Deliverance docked at Ordnance Island, St. George's. Compare its size to the small yacht on the right.

ORDNANCE ISLAND:

When you are ready, retrace your steps to King's Square. Proceed to the far side of the Square, where a replica of the Deliverance is docked at Ordnance Island. As described earlier in the tour, after initial construction at Building Bay, the ship was brought to this area for completion. Governor Gates and almost a hundred of the colonists continued their voyage to Jamestown on this pinnace. The size of the second boat, Patience, can be gauged by imagining a craft about half the size of the Deliverance. Sir George's journeys in that small vessel eventually covered some 2,300 miles before he returned to Bermuda.

Located in King's Square today, the recreations of stocks and "ducking" are popular with modern visitors; they were, in fact, some of the less severe punishments performed in the colony. Although known today as Ordnance Island, this area was originally composed of two smaller islands used for those punishments:

Ducking Stool and Gallows Islands. As the last name implies, Gallows Island was used for executions for crimes as simple as stealing food. The shallows and original islands were built up by the British Royal Army in the 1800's to store munitions.

Located in the park across the road from the Deliverance is the statue of Admiral George Somers. Sculpted by Desmond Fountain, the piece was dedicated by HRH Princess Margaret on the 375th Anniversary of Bermuda in 1984.

Find a comfortable spot and remain in this area for the following text.

Finally, despite their misgivings, the investors sent a relief ship; laden with the needed supplies, the *Welcome* reached the colony early in 1615. Yet, although Moore's three-year commission as governor was nearly at an end, to the Governor's surprise (and probably relief), the ship carried no word from London about a new appointment. As Captain John Smith recounted soon after:

Master Moore seeing they sent not for him, his time being near expired, understanding how badly they reputed him in England, and that his employment now was more for their own ends then any good for himself, resolved directly to return with this ship...

After spending three years setting the colony at Somers Isles on its feet, Richard Moore gathered up his family and belongings and sailed back to England with little fanfare. Some 225 years later, another British historian would off-handedly grant Richard Moore his due: "Governor Moore was a man of ordinary condition, being a carpenter, but in every respect he showed the prudence of the choice of the

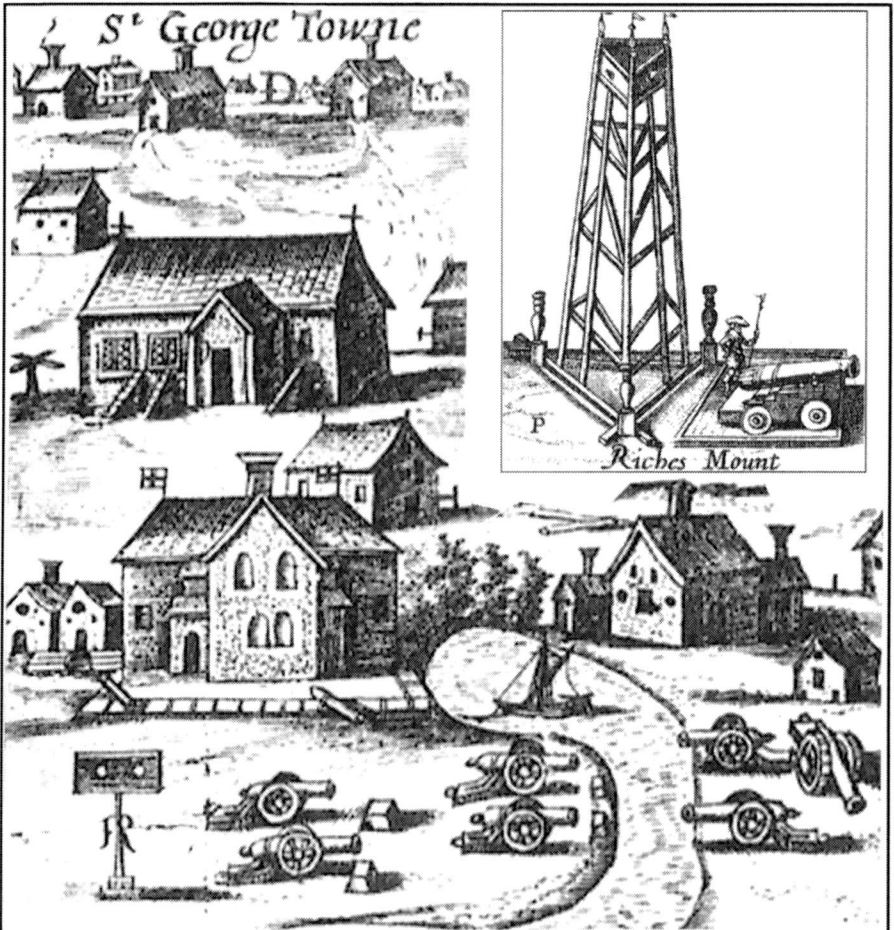

Town of St. George as it appeared by 1622. Details include: stocks in the Town Square, with Government House on Water Street just behind, St. Peter's Church on the hillside above, and the "Riches Mount" watchtower (inset at right). (Smith, "General Historie")

Proprietors, who owed every thing to his sagacity, firmness, and prudence." 26

1615: The Somers Isles Company

Despite their difficulties, the colony at Somers Isles under Moore's administration had been modestly successful. However, and at perhaps the worst possible time, the Virginia Company in London was struggling with internal problems. As early visions of a fortune in pearls and ambergris faded, it became clear that further profits from agriculture or land sales would be limited on the tiny archipelago. Splits within the Company soon developed over where those limited profits would be spent. In the end, 117 investors still believing in the colony purchased the rights to Somers Isles for £2000. On June 29, 1615, King James issued a new charter; the "London Company of the Somers Isles" would oversee the settlement with new energy and a common purpose.

Yet even as the governance in London reorganized, on Somers Isles the leadership faltered. When the Company had chosen no one to replace him, before leaving the islands, Moore himself chose a six-man council of deputy-governors; each man was to serve for a month until the position could be filled by an appointment from London.

Described with perhaps faint praise as, "the best that were there," Moore's choices included two long-time settlers from *Sea Venture,* Edward Waters and Christopher Carter, as well as Captain Miles Kendall, Captain John Mansfield, Thomas Knight, and Charles Caldicot. It appears none of these six men apparently shared Moore's vision for the settlement. Sadly, after all that had been accomplished, for several months Somers Isles saw a period of misrule, contention, and a "time of ease"; as no one was directed to work, not surprisingly the development of the colony ground to a halt. If Richard Moore had put the colony on its feet, it would take the next appointed Governor, Daniel Tucker, to get it up and running again. 27

Epilogue

By early 1616, there were no shipments of tobacco (much less ambergris) reaching London; in their stead, the Company was receiving reports of "disgraceful revelry" on Somers Isles. Finally, to restore order in the unproductive colony, that March the Somers Isles Company appointed Daniel Tucker as governor. Known more for his temper than his judgment, what Tucker lacked in congeniality he made up for in experience. While serving as cape merchant for Jamestown, Tucker learned all too well that successful crops and obedience to law were essential to the survival of a colony, In short, under Tucker's leadership, the colonists would have to work to eat.

Largely despised by the colonists, particularly after the previous lax administration, Tucker comes down to us as a tyrant. Yet the colony started to produce; within a year, the rat population was under control, fishing resumed around the islands, and crops of corn, tobacco, as well as new fields of fruit and sugar cane were indeed thriving. The following year, a shipment of 30,000 pounds of the colony's next profitable commodity—tobacco—arrived in England.

That year, Governor Tucker also commissioned Richard Norwood to complete his survey: the eight equally sized "tribes" plotted by the surveyor were divided into fifty twenty-five acre shares for Company investors. (Interestingly, the precise boundaries plotted by Norwood have been used to define property to this day).

Also in 1616, there were new arrivals at Somers Isles that would change the course of its society for years to come. Still hoping to harvest a non-existent fortune in pearls, the Company brought in two slaves, a black and an Indian, from the West Indies. When the search

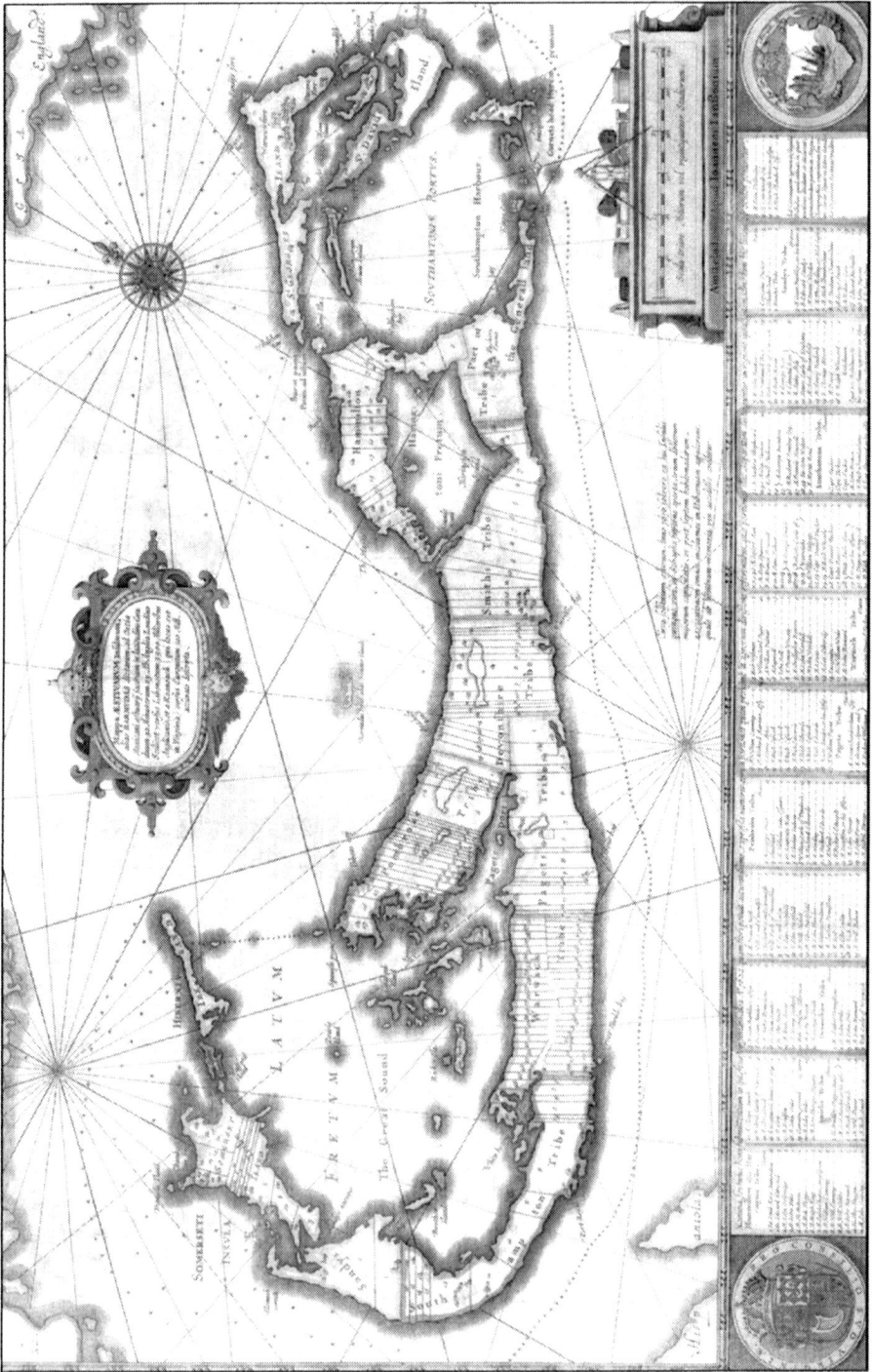

Richard Norwood's Survey as depicted in Blaeu's 1635 engraving.

again proved fruitless, the pair were put to work harvesting crops of tobacco and sugar cane. Later that year, a privateer vessel arrived, trading fourteen captured Blacks for supplies. In the end, the Company system of indentured workers and small farms would never require the large numbers of slaves used in other English colonies; nonetheless, the importation of West Indian and African slaves created a system of racial inequality that would haunt Bermuda for hundreds of years to come. 28

Even during Tucker's more prosperous administration, the Company's distant London offices continued to send demands: the shareholders were still clamoring for a return on their investment. The potential profit from tobacco notwithstanding, Daniel Tucker's concentration on crops made him "fitter to be a gardener than a governor." On the expiration of his three year commission, Tucker was also not reappointed, and Nathaniel Butler became the next governor of the islands.

However, although London continued to appoint governors, by 1620, the colony gained a measure of self-government when a colonial parliament was created, the House of Assembly. (One of the bills to be passed by the Assembly protected the shrinking cahow population—the first conservation measure of its kind in the New World).

Lining the neat streets of the Town of St. George and dotting the islands beyond were the shops, homes, and farms of nearly a thousand settlers; with the islands' healthy climate and low mortality rates, that population grew rapidly. By 1617, the small patches of tobacco discovered by *Sea Venture* survivors eight years before had been expanded, and for a time became the islands' only principal, and profitable, export. 29

The defenses started under Governor Moore were enlarged, and the islands' approaches were now defended by eight substantial forts. The "religious government" initiated under Moore now embraced some six churches across the islands. In short, by 1620, the Somers Isles were on their way to be a model colony.

In the end, the fact that these reef-bound islands came to be settled at all, much less prospered, is remarkable. Chance occurrence or no, when the frightened survivors of the *Sea Venture* stumbled ashore in 1609, their tenacity, along with the leadership of Thomas Gates and the vision of George Somers, set in motion events that went far beyond all expectation.

"Blessed bee God that suffered Sir Thomas Gates and Sir George Somers to be cast away upon these islands…

- Richard Moore, 1613

This stop marks the end of the formal tour; I hope you have enjoyed this walk into Bermuda's past. While you are in the Town of St. George, a sampling of other historic sites near King's Square includes:

- Town Hall: Located on the east side of the Square, the distinctive limestone structure dates from 1782. Like St. Peter's Church, it has beautiful cedar floors and ceilings. The St. George's civic government still meets in this building.

- Bermuda National Trust Museum: Located on Duke of York St. across from St. Peter's Church. Built as a Government House by Governor Samuel Day in 1699, it was later used by an agent of the Confederacy during the American Civil War. (Of interest to this tour, there is a model of the ship Sea Venture on the first floor).

- St. George's Historical Society Museum: Located two blocks north of the square on Duke of Kent Street. Built in the 1730's, this is one of the oldest buildings in Bermuda. The museum has exhibits and collections reflecting day to day life in 1700s.

- Bermudian Heritage Museum: Located west of the square on Water Street. Dedicated to interpreting the islands' Black History, its exhibits reflect the historical heritage and achievements of the islands' black population.

- Tucker House Museum: Located west of the square on Water Street. Built in 1719, it was eventually home to Bermuda's Tucker family and another link to Virginia. Artifacts and collections of cedar, china, and silver describe their day to day lives.

- World Heritage Centre: Located next to Penno's Wharf Cruise Terminal at the western end of Water Street. Opened in 2009, the center offers an overview of the town, its heritage, through a diorama, displays, and videos.

- Also in Bermuda at the Royal Naval Dockyard/King's Wharf:
National Museum of Bermuda: (formerly the Maritime Museum) Expanded on Bermuda's 400th Anniversary, displays and programs now showcase the islands' largest collection of artifacts reflecting all of Bermuda's cultural, maritime, and military history.

Appendix

A. Bermuda's North Rock

So significant in the founding of Bermuda, the reefs that virtually surround the islands have long fascinated scientists. The archipelago of Bermuda today is actually a seamount formed by the high points on the rim of a submarine volcano. Over the millennia, the top of this seamount has seen alternating lives both above and below sea level. When vast amounts of seawater were frozen in the icecaps during the Ice Age, Bermuda existed as a landmass that likely extended from the reef terrace nine miles north of Flatts Village, to the shoals off South Shore, to those reefs beyond the east and west shores—an island of some two hundred square miles.

Rarely seen, much less visited by tourists, is the sole remnant of what was once the northern rim of that island. Up through the nineteenth century, North Rock was composed of at least five

North Rock Flat as it appeared in 1875

limestone boulders, the tallest standing some fourteen feet above the waves. Over time, the coral flats that spread beyond North Rock became the graveyard of numerous ships, notably Englishman Henry May's 1593 wreck.

And, as possibly the oldest exposed rock formation in Bermuda, the unusual site gained the attention of early scientists. Nineteenth-century naturalist Angelo Heilprin described them as, "beyond comparison, the most interesting feature which the Bermudas have to offer." Professor Heilprin may be forgiven what we today would view as an overstatement, for North Rock at that point was a stunning formation, seemingly far out to sea, and a broad flat you could walk on at low tide.

Although still iconic, the formation of North Rock has changed dramatically. Today, with the rise of sea level—about two feet since Henry May's time, the incessant pounding of wind and waves, and possibly some unfortunate target practice by the military, only two boulders remain above the waterline: the larger pillar, still standing about twelve feet tall, and a smaller boulder. Dwarfing the formation today, a seventy-foot tall navigational beacon stands to warn modern mariners from that broad reef where so many others have come to misfortune.

The Shipwreck on Bermuda's Coat of Arms

Exploring the once prominent formation at North Rock often leads to an interesting conjecture, and one worth exploring. Prominently located on the right side of Bermuda's flag, one notices the official Coat of Arms; the formal blazon specifies: "Argent, on a mount vert a lion sejant affronté gules supporting between the fore-paws an antique shield azure thereon a representation of the wreck of the ship *Sea Venture* proper." (Or without the formal language of heraldry: "On a bright white background, on a grassy mound, the red lion of England faces front sitting on its haunches, holding a blue shield picturing the wreck of the *Sea Venture*). Considering the story contained on these pages, the fact that Bermuda's seal carries the image of a shipwreck should come as no surprise; what may be a surprise, however, is that the source of the image may not represent the wreck of the *Sea Venture*.

Formally adopted during the 300th Anniversary of Bermuda's founding in 1910, this coat of arms replaces a badge in use since the early 1800's depicting three sailing ships and a dock, symbolizing the Islands' maritime heritage. Doubtless more dramatic, the new seal was based on an image that first appeared in 1624 in John Smith's Generall Historie of Virginia, New-England, and the Summer Isles.

Close-ups of: Norwood's Crest, c. 1622 (l), Smith's Crest, 1624 (c), North Rock, 1875 (r)

However, it is fairly certain that Smith had copied the island map and crest from Richard Norwood's previous survey maps. 30

On his map, Norwood, drew the Virginia Company seal in one corner, but on the other, a basic version of a Somers Isles crest (doubtless on Company specification); while similar, the crest is far from what today's blazon specifies (see image). The 1910 interpretation of the crest as the famous *Sea Venture* wreck was a fitting addition to the Anniversary. However, Norwood was an excellent draftsman, and even allowing artistic license, he clearly depicts a shipwreck occurring on a vertical shelf of boulders, not a submerged coral reef. One might well suspect this image was intended to accurately represent the wreck of Henry May in 1593— the first Briton of record on the islands, and ten years before Spain's Captain Ramirez—not the arrival of Sir George Somers and the survivors on the *Sea Venture* sixteen years later.

(For more information, see *Look Bermuda*'s Vimeo project, 'The Riddle of the Crest," featuring research by Dr. Philippe Rouja, Principal Scientist of Bermuda's Department of Conservation Services).

B. Where Is the Sea Venture?

Almost as soon as the Sea Venture slipped beneath the surf off Bermuda's East End, there was some question as to where she went down. An experienced mariner who certainly should have known the location, Sir George Somers, wrote days later that the ship was "a quarter of a mile distant from the shoare." Passenger William Strachey's otherwise accurate account claimed, "three quarters of a mile," and Sylvester Jourdain described, "half an English mile."

Bermuda's East End, Lemrpriere Sea Chart, 1826

Although up through 1622, items were retrieved from the wreck site, by 1826 the area known as "Sea Venture Flat" covered an area of over a square mile. The issue lay dormant for 120 years until 1958, when an amateur skin diver (and descendent of one of the Sea Venture's survivors) named Edmund Downing located a decayed hull a mile from the shore. Aided by Bermuda wreck expert Teddy

Tucker, the pair located a number of period artifacts near the surviving timbers, including the barrel of a small cannon. Coinciding with the 350th Anniversary the following year, the find stirred the international interest and funding from the Bermuda Government— until a London expert identified the piece of ordnance as a later Spanish cannon. Funding for the investigation dried up, and the matter was dropped for another 20 years.

Finally, in 1978, another diver alerted the newly formed Bermuda Maritime Museum to the site, funding was obtained, and the investigation began anew. Within two years, the site was uncovered using modern underwater equipment, revealing the remaining keel, ribs, and a trove of period artifacts, all identified as consistent with an English ship of 1609. This data led to a reassessment of the cannon barrel, and after twenty years, Edmund Downing's site was finally confirmed as being that of the *Sea Venture* (see map of approximate location on page 37). In the course of the research, the remaining previously buried timbers were mapped; combining the map with digitized drawings of 1600's ship-building techniques has allowed researchers to model an accurate idea of the size and appearance of the *Sea Venture*'s hull. More recently, with the help of the Bermuda Regiment, the surviving timbers have been re-buried to prevent further deterioration.

C. Sea Venture Passengers

Of 150 survivors of the wreck of the *Sea Venture*, the names of only fifty are confirmed and engraved as below on the *Sea Venture* monument. Additional information from other sources has been added in parentheses.

Virginia Company Officers

Lt. Gen. Sir Thomas Gates (*Governor for Virginia*)
Admiral Sir George Somers (*Admiral of Third Supply fleet*)
Captain Christopher Newport (*Captain of the Sea Venture*)
Reverend Richard Buck (*Chaplain, also spelled Bucke*)
and
Mistress Maria Thorowgood Buck (*With possibly two daughters*)
Thomas Whittingham *Killed by Native People in Virginia
(*Cape Merchant*)

Adventurers

Ralph Hamor, Esq.
Mistress Horton
Silvester Jourdain, Esq. (*of Lyme Regis, Dorset*)
Henry Paine, Esq. * Executed (*Mutineer*)
Capt. Wm. Pierce
Robert Rich, Esq.
Samuel Sharpe, Esq.
Henry Shelley, Esq.
William Strachey, Esq.
James Swift, Esq.
Capt. George Yeardley, Gates Aide

Colonists

Henry Bagwell
Humphrey Blunt
Jeffrey Briars *Died on Bermuda
Joshua Chard (*Or Joseph*)
Edward Eason &
Mistress Eason ** Infant son "Bermudas" born in 1610
George Graves
Thomas Godby
William Hitchman *Died on Bermuda

Stephen Hopkins
Elizabeth Jones (*Or Joons, aged 30, servant*)

Richard Lewis *Died on Bermuda
John Lightfoot
Elizabeth Persons, (*Or Parsons, maid to Mistress Horton - married
Thomas Powell in Bermuda*)
John Proctor
Humfrey Reede
John Rolfe &
Mistress Rolfe *** Infant girl "Bermuda" born and died on Bermuda
(*first wife of John Rolfe*)

Mariners

Edward Waters, Somers Aide (Lieutenant - *remained on Bermuda
1610-?*)
Henry Ravens, Master mate, **Killed by Native People in Virginia
(*Lost w/six unnamed crewmen*)
Robert Walsingham, Coxswain
Robert Frobisher, Shipwright
Nicholas Bennit, Carpenter
Thomas Powell, Cook (*Married Elizabeth Persons in Bermuda*)
William Brian
Christopher Carter **Stayed Behind in Bermuda (*Mutineer*)
Edward Chard (*Mutineer- Remained on Bermuda 1610-?*)
Richard Knowles
Francis Pearepoint
Edward Samuel, ***Murdered (*by Robert Waters*)
John Want
Robert Waters **Stayed Behind in Bermuda (*Killed E. Samuel;
Mutineer, remained on Bermuda 1610, returned to England 1610*)

Apparently considered unimportant to the colony, other than the women named, records of the other females on the journey are not extant.

In addition, according to Captain Smith's "Historie of Virginia," *There were two Savages...one called Namuntack, the other Matchumps,....such differences fell between them, that Matchumps slew Namuntack, and ...made a hole to bury him...which murder he concealed till he was in Virginia.*

Although the accuracy of some of Smith's writing has been called into question, later accounts appear to corroborate the existence of two Native Americans. Apparently first brought to England during the early Jamestown settlement, Namuntack, the son of a Powhatan chief, and a servant were returning to Virginia aboard the *Sea Venture* when the murder occurred. Similar to the records of the *Sea Venture's* female passengers, the fact that no one else reported the murder may well be evidence of how little esteem Native Americans held.

Endnotes

1. The Cahow (more properly the Bermuda Petrel), is an endemic species named for its strange cry. The fowl and its eggs were overhunted as an early food source, and the bird was thought to be extinct for over 300 years. Amazingly, a few nesting pairs were discovered in 1951; the success of fifty years of conservation projects and legal protection has gradually restored some colonies on Nonesuch Island.

2. Known for years as Spanish Rock, modern research indicates the inscription was probably made by marooned Portuguese sailors. The more recent interpretation of the letters is "RP" for Rex Portugaline, or, King of Portugal.

3. Addison E. Verrill, The Bermuda Islands, (New Haven: Tuttle, Moorehouse), pp. 198-204.

4. Verrill, pp. 122-4. As the first Briton to land on the islands, in effect "claiming" them for England, May and his account were well known and appear in many early histories. According to Verrill, May's wreck at North Rock was the basis for the first Bermuda seal (see Appendix A).

5. Ibid. For almost ten years, the Virginia Company's optimism over finding pearls seems to have overridden the reality.

6. The "tunne" (or tun) was a term representing the volume in cargo capacity, not the weight of a given ship.

7. Jean Kennedy, Isle of Devils: Bermuda under the Somers Island Company, 1609-1685, (New York: Collins) pp. 21-2.

8. John Smith, John, The Generall Historie of Virginia, New-England, and the Summer Isles, Vol. 1 (London: Printed by I.D. and I.H. for Michael Sparkes), pg.90. In the apologetic, "True Declaration of the Estate of the Colonies in Virginia," Gates demurred, making the distinction that there were no other previously appointed "commissioners" aboard the Sea Venture, and so leadership was not an issue.

9. The Company of Merchant Adventurers, a trading guild that gave rise to the Virginia Company, owned a number of ships. As a

founding member of the Company, it seems likely Somers had ownership in a number of those vessels, including at least part of the *Sea Venture*. The dates for the age of the ship vary, although Jonathan Adams' research in <u>Maritime Archaeology</u> points to the 1603 date.

10. Robert Waters and Edward Waters are frequently confused in Bermuda histories; both were both part of the *Sea Venture*'s crew. However, Lt. Edward Waters was Admiral Somers' aide, and remained on Bermuda in 1610 after Somers' death (also see *Note 27*). Robert Waters murdered Edward Samuel on Bermuda, and also became a mutineer. He remained behind on Bermuda in 1609 to avoid punishment, but returned to England in 1610.

11. William Strachey, <u>True Reportory of the Wreck and Redemption of Sir Thomas Gates, Knight</u>, in <u>A Voyage to Virginia In 1609: Two Narratives</u>, Ed. Louis B. Wright, (Charlottesville: Univ. of Virginia Press). Strachey's "True Reportory" was originally a letter to an investor of the London Company. Doubtless Strachey knew the letter would be circulated. All of Strachey's quotes contained herein are from this account.

12. Sylvester Jourdain, <u>A discovery of the Bermudas, now called the Sommer Hands, 1609-1610</u>, in <u>A Voyage to Virginia In 1609: Two Narratives</u>, Ed. Louis B. Wright, (Charlottesville: Univ. of Virginia Press). All of Jourdain's quotes contained herein are from this account.

13. The "whipstaff" was an extension of the tiller used for larger vessels in the days before the ship's wheel was developed. The helmsman moved the lever located in a cabin below the sterncastle; having a rather limited view forward through the port in the deck above, he relied largely on relayed directions.

14. Some later accounts such as Stow's 1632 <u>General Chronicle of England</u> claim that Somers, Newport, and Gates agreed before the wreck to try to land despite the islands' evil reputation. Considering the conditions existing on the ship in the hour before the landing, and the fact that Stow's history is loosely based on other accounts, such a conference seems unlikely.

15. Want's other conspirators were Christopher Carter, Francis

Pearepoint, William Brian, William Martin, and Richard Knowles. The information we have on these mutinies comes from Strachey's account. Oddly, neither Jourdain's account nor Somers' letter to the Company mention the mutinies.

16. Due to Somers' explorations, the largest of the islands was termed the "main island" (as it is today). There is no record of where Somers' camp was placed, however, a site at the cove at Walsingham Bay in Hamilton Parish has the necessary features and is a popular theory.

In the rift between William Gates and George Somers, it is interesting to speculate where Captain Christopher Newport's allegiance lay: he was a fellow mariner to Somers, but considering his previous experience managing personality clashes at Jamestown, he could have ended up in either camp.

17. According to Tucker's well-known history, Bermuda, Today and Yesterday, Stephen Hopkins acted "in exactly a similar fashion" on reaching America. This author finds no evidence of this behavior in Jamestown. However, Hopkins is probably the same colonist who later travels on the *Mayflower*. In Plymouth, he eventually had a brush with the law for assault and his management of a tavern; but more importantly, he was a positive influence in development of the colony and its relations with the Native American peoples.

18. Verrill's history claims the "little round island" was one of the small islands that later formed "Ordnance Island" on the shore of St. George's.

19. To prepare supplies for Virginia would have taken time, and Somers became ill soon after his return to Bermuda; by November, he was still on the Islands. Theories on the reasons for Somers' actions after Jamestown run the gamut: a humanitarian gesture, his desire to establish Bermuda as a colony, a conspiracy with Carter and Waters, or simply, his outright desire to get away from the conditions at Jamestown. In any event, his offer to obtain supplies was not outside of his responsibilities to the Company or his personality.

20. "The small amount of sail" refers to Strachey's mention that the *Patience* had to borrow sails to keep up with *Deliverance* on the trip to Jamestown.

21. William Williams, <u>An Historical and Statistical Account of the Bermudas From Their Discovery to the Present Time</u>, Vol. 1, (London: Thomas Cautley Newby, Publ.) 1848, pp.16-17. As Matthew Somers did not return to Virginia as duty required, it seems likely he returned the Admiral's remains to England to avoid questions—and establish a possible claim to the Admiral's estate (Matthew would receive nothing). Somers was buried near his home at Whitchurch, in Dorsetshire. If, as some claimed, the Admiral's last wishes were to be buried on the islands, at least his nephew obeyed in part.

The spot where Matthew Somers buried his uncle's heart was marked with a wooden cross. In 1619, Governor Nathaniel Butler marked the spot with a marble slab that read:

In the yeere 1 6 1 1 Noble Sir George Summers went hence to heaven, Whose well-tried worth that held him still imploid Gave him the knowledge of the world so wide; Hence 'twas by Heaven's decree that to this place He brought new guests and name to mutual grace; At last his soul and body being to part, He here bequeathed his entrails and his heart.

Curiously, in 1819, Admiral Sir David Milne and some St. George's officials decided to open the vault (and an adjacent 1726 interment); they found only part of an opaque glass bottle and some decayed bones (Sir George's remains had probably long since turned to dust). According to Williams' 1848 account,

The two monuments were very roughly handled at the time and left in a more dilapidated state than when disturbed, it is questionable if this violation of the sanctity of the grave can be termed anything better than wanton curiosity. A subscription was projected at the time for enclosing the place, and erecting a suitable memorial; but it was never done, and the broken stones remain a reproach to the authorities.

The approximate site was marked with the existing marker by Governor John Lefroy in 1878. Due to the widening of Duke of York Street in the past 400 years, it seems likely that the burial site is under the present roadbed.

22. Raised in poverty, Richard Norwood developed a taste for mathematics early in life, and soon became something of a genius at navigation and surveying. He eventually completed the Bermuda survey in 1617 under Governor Tucker, and various engravings of it were printed after 1624.

23. Although primitive, Moore's defenses were sited well and became the forerunner of many of the British forts that ring Bermuda today.

24. Although not known until years later, Spain's King Phillip III had finally become interested in Bermuda as a way-station and ordered ships to investigate the English colony reported to be there. The results of this probe once again deterred the King from making claims to the Islands.

25. The colonists sent to Virginia in 1607 were similarly unprepared. However, at Jamestown, it was the London Company's search for gold that diverted settlers from agriculture and other work necessary to maintain the colony.

26. After living in poverty for a time in London, Moore was eventually given some of the recognition he deserved; the new Somers Island Company awarded him eight shares of land, and, "so he was dismissed of his charge, with show of favour and much friendship." Richard Moore died in 1618.

27. Almost immediately, the leadership had problems: Caldicot devised a plan with Knight and Waters to relieve the colony's food shortage; they would sail to the Indies to gather supplies. After a series of calamities, including being blown off course to the Canaries, robbed by French pirates, then shipwrecked and saved by English pirates, most of these mariners never returned to Bermuda. Waters may have made it back to the Islands, but by 1616 moved to Virginia. In the census of 1623, Waters apparently owned 100 acres near what is now Warwick, VA. Waters, his wife, and son are listed at Elizabeth City. At one point, the couple were taken prisoners and held by the Nansemond Indians, but eventually escaped. Christopher Carter remained in Bermuda after his brief posting as Governor, however he was killed in an accident in 1623.

28. There was a fine line between slavery and "indentured worker," particularly later when the period of indenture for West Indians was increased from seven to 99 years. Those enslaved also came to include prisoners captured in England's conflicts: Native Americans from Virginia, as well as prisoners from Ireland and Scotland.

29. The islands' tobacco crop became profitable for a time (and was even used as currency), but eventually could not compete with the quality and quantity of tobacco produced in Virginia. "Tobacco was exposed to competition from so many quarters," wrote Governor Lefroy, "that its value must have rapidly declined without the crushing effect of the taxes upon it."

30. Norwood's first map of Bermuda was completed in 1618, and clearly is the basis for John Smith's map. In 1622, Norwood registered another survey map in London records. Although no copy of this map survives, Speed's 1626 engraving of Norwood's map is apparently based on this version.

Bibliography

Books

Cooper, Frederic T., ed. Fremont Rider, <u>Rider's Bermuda: A Guide Book for Travelers</u>, (New York: Henry Holt and Company) 1922

Council of Virginia, <u>A True Declaration of the Estate of the Colonies in Virginia</u>, (London: W. Barret) 1610, Electronic Edition: Furman University Digital Collection, 2010

Dean, James S., <u>Tropic Suns: Seadogs Aboard an English Galleon</u>, (New York: The History Press) 2014

Doherty, Kieran, <u>Sea Venture: Shipwreck, Survival, and the Salvation of Jamestown</u>, (New York, St. Martin's Pr.) 2007

Glover, Lorri & Smith, Daniel Blake <u>The Shipwreck That Saved Jamestown: The Sea Venture Castaways and the Fate of America</u>, (New York: Henry Holt & Co.) 2008

Heilprin, Angelo, <u>The Bermuda Islands: A Contribution To The Physical History And Zoology Of The Somers Archipelago</u>, (Philadelphia: Publ. by Author) 1889

Jourdain, Sylvester, <u>A discovery of the Bermudas, now called the Sommer Hands, 1609-1610</u>, in <u>A Voyage to Virginia In 1609: Two Narratives</u>, Ed. Louis B. Wright, (Charlottesville: Univ. of Virginia Press) 2nd Ed., 2013

Kennedy, Jean, <u>Isle of Devils: Bermuda under the Somers Island Company, 1609-1685</u>, (New York: Collins) 1971

Lefroy, John H., <u>Discovery and Early Settlement of the Bermudas or Somers Islands Vol.1</u>, (1515-1582), (London: Longmans, Green, and Co.) 1877

Mattingly, Garrett, The Armada, (Boston: Houghton Mifflin) 1959

Murray, Hugh, An Historical and Descriptive Account of British America In Two Volumes, Vol. II, (New York: Harper & Brothers) 1840

Smith, John, The Generall Historie of Virginia, New-England, and the Summer Isles: 1584 To This Present 1624, (London: Printed by I.D. and I.H. for Michael Sparkes), 1624. Electronic Edition: University of North Carolina, Chapel Hill, 2006.

Stow, John, & ed., Howes, Edmund, Annales, or a General Chronicle of England, (London: Richard Meighen) 1632; Electronic Edition: Google Books/WorldCat

Strachey, William, A True Reportory of the Wreck and Redemption of Sir Thomas Gates, Knight, in A Voyage to Virginia, Ed. Louis B. Wright, (Charlottesville: Univ. of Virginia Press) 2nd Edition, 2013

Tucker, Terry, Bermuda, Today and Yesterday, 1503-1980's, (London: Robert Hale, Ltd) 2nd Ed., 1983

Verrill, Addison E., The Bermuda Islands, An Account of Scenery, Climate, Etc., (New Haven: Tuttle, Moorehouse) 1903

Whymper, Frederick, The Sea: Its Stirring Story of Adventure, Peril & Heroism , (London: Cassell, Limited) 1883

Williams, William, An Historical and Statistical Account of the Bermudas From Their Discovery to the Present Time, Vol. 1, (London: Thomas Cautley Newby, Publ.) 1848

Papers, Articles & Letters

Adams, Jonathan, A Maritime Archaeology of Ships: Innovation and

<u>Social Change in Late Medieval and Early Modern Europe</u>, (Oxford: Oxbow Books) Electronic Edition: Google Library, 2013

Andrews, Charlotte, "Community Uses of Maritime Heritage in Bermuda," Doctoral Dissertation in Philosophy, University of Cambridge, 2010

Kennedy, Neil, "William Crashaw's Bridge: Bermuda and the Origins of the English Atlantic." Essay from: Nancy L. Rhoden, <u>Revisiting The English Atlantic in the Context of Atlantic History</u>, McGill-Queen's University Press, 2007, Pg. 107-135.

Somers, George, Letter to the Earl of Salisbury, Jane 20, 1610. (Reprinted in J.H. Lefroy, <u>Discovery and Early Settlement of the Bermudas or Somers Islands Vol.1</u>, Pg. 10-11.

Thomas, Martin, "A Teaching Guide to the Biology and Geology of Bermuda" March, 2006 Publ. by Bermuda Zoological Society in Collaboration w/ Bermuda Aquarium, Electronic PDF Edition courtesy of BAMZ

Websites:

Bermuda.Com (Definitive Guide to Bermuda)
www.bermuda.com/
Bermuda Online (A link to all things Bermuda: 145 websites)
www.bermuda-online.org/
Bermuda Department of Education,
"Bermuda: Five Centuries"
"St George's; History and Select Buildings"
www.moed.bm/default.aspx
The Bermudian, "The Wreck of the Sea Venture: The Untold Story"
/www.thebermudian.com/heritage/
Bernews, (Online) Sea Venture Memorial
bernews.com/2010/11/photos-sea-venture-memorial-unveiled/

Encyclopedia Virginia, (Biographies & related information)
www.encyclopediavirginia.org/

Government of Bermuda, (Maps & other resources)
www.bermudamaps.bm/

Look Bermuda Destination Media
"Downing's Wreck; the Story of the Sea Venture"   .
"'Higher Ground' the Cahow Translocation Project Film"
www.lookbermuda.com/home/

National Humanities Center, American Beginnings: The European Presence in America 1492-1690
nationalhumanitiescenter.org/pds/amerbegin/index.htm

The Royal Gazette (Online) Heritage Matters Series by Dr. Edward Harris, Director, National Museum of Bermuda
www.royalgazette.com/

Index

Notes

About the Author

Although an historian of the American Civil War, John Archer has long been fascinated with Bermuda and its history. In addition to writing, he offers tours as a Licensed Guide for Gettysburg National Military Park. He has conducted several of the Park's Anniversary "Battlewalk" series, and the tours featured on Pennsylvania Cable TV.

The author's other written work includes three interpretive tour books of various parts of the battle, as well as articles in several history periodicals. His first work of historical fiction, "After the Rain: A Novel of War and Coming Home," received Director's Mention for the Langum Prize in American Historical Fiction. This is the author's first book on Bermuda.

CPSIA information can be obtained
at www.ICGtesting.com
Printed in the USA
LVOW10s0228220517
535366LV00008B/506/P